VISUAL GALAXY

VISUAL GALAXY

The Ultimate Guide
to the Milky Way and Beyond

NATIONAL
GEOGRAPHIC

Washington, D.C.

THE TUMULTUOUS ORIGIN OF THE MILKY WAY

Like many galaxies, our own, the Milky Way, got its current size and shape by colliding and merging with other galaxies. In the aftermath, wherever the concentration of gas and dust was higher, thousands of new stars lit up the universe.

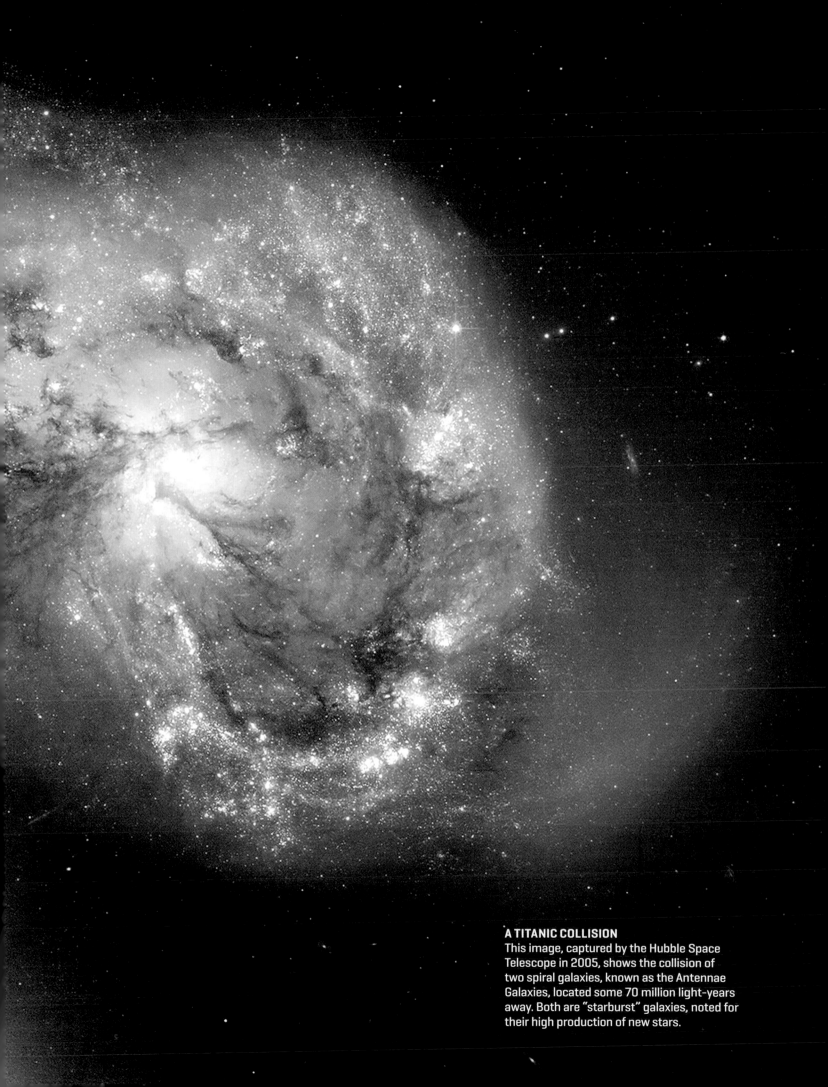

A TITANIC COLLISION
This image, captured by the Hubble Space Telescope in 2005, shows the collision of two spiral galaxies, known as the Antennae Galaxies, located some 70 million light-years away. Both are "starburst" galaxies, noted for their high production of new stars.

THE TWO SIDES OF THE SUN

Some 4.6 billion years ago, the star that governs our solar system and provides all our energy was born in the Orion Arm of the Milky Way. The eventual transformation of the sun into a red giant will mark the end of our solar system.

THE SUN'S TURBULENT SURFACE
A sunspot, like the one seen in an image captured by the Swedish Solar Telescope (SST), can reach several times the size of Earth. Its magnetic activity is so intense that it can cause solar storms.

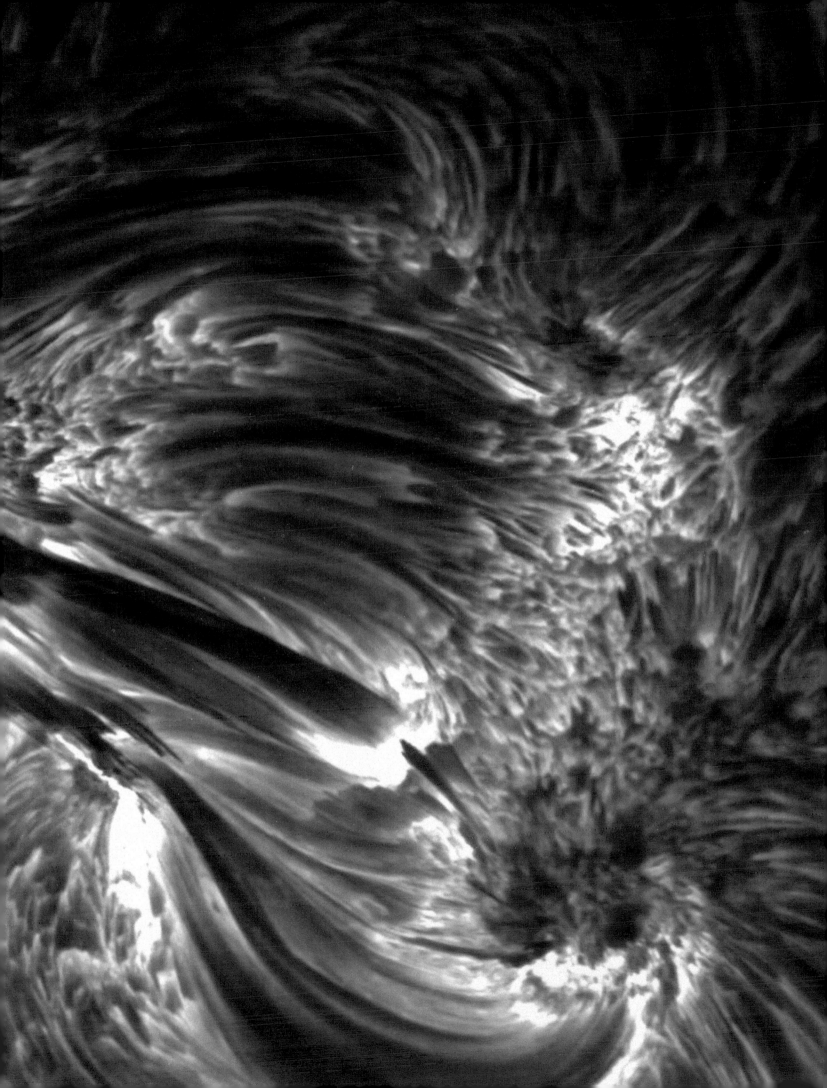

BIRTH OF THE MOON
This illustration imagines the impact by an object the size of Mars on a young planet Earth. Most of the object was absorbed by the planet; the remaining debris formed our moon.

THE PLANETS,
A HISTORY OF VIOLENCE

The planets are children of chaos. When the sun was just a few
tens of millions of years old, fragments of rock and ice orbiting
it collided repeatedly and ended up fusing into planets.

THE EMERGENCE OF LIFE

Once the planetary orbits settled, the solar system became stable. The first trace of life on Earth appeared a little later, some 3.5 billion years ago.

EARTH TODAY
The characteristics that make life on Earth possible are unique in the solar system, although we don't know if those same principles hold true for the rest of the universe. It is estimated that billions of solar systems like ours exist in the Milky Way alone.

CONTENTS

FOREWORD

Col. Chris Hadfield
Astronaut and Former Commander
of the International Space Station

Longer ago than anyone can remember, far before all written history, we were explorers. The lasting remains of our early ancestors—their lost stone tools, their doused campfires, and their scattered, fossilized bones—are spread across the Earth. Right from the beginning, from central Africa to Australia to the Andes, all the way around the planet, our traces show that we have never been satisfied to stay in one place.

That innate wanderlust starts from birth, programmed into our fundamental human genetics. Watching a toddler grow, you can see the tiny explorer at work. We develop the ability to walk at around 12 months, a full year or more before we can talk, before anyone can clearly explain things to us. Our burning desire to take those first wobbly steps is rooted in our need to go see for ourselves, to taste and touch the world around us. We learn by exploring.

But we *Homo sapiens* are very limited in physical ability. Many animals are faster runners and better climbers. Our eyes are made mostly for daylight, seeing poorly at night, making us hide fearfully when it's dark. We can teach ourselves to swim, but neither far nor fast, forcing us to stay close to shore. Our fur is too thin to protect us from cold, and our soft feet don't do well on sharp rocks. Our fangs and claws are unimpressive, scaring no predator. We have no wings to fly with. These human bodies that our DNA builds are not natural exploring machines.

What we do have, though, is imagination. It's perhaps what best defines us. We can see a common rock and picture it as a hammer, or if broken, a blade. We long ago realized that we could wear other animals' skins for warmth, and that we could use the plants and rocks and even the snow around us to build shelter. We lashed logs together and became sailors. We watched lightning strike, felt the heat, and taught ourselves how to control fire. And as humans we began to look to the horizon and imagine what might lie beyond it.

But when we raised our eyes above the horizon, our imaginations were truly unleashed. Our ancestors looked to the sky, to the sun and moon, and created amazing stories of what those two things were and how they moved. In our minds the sun became a flaming chariot driven by the god Helios. The moon was a daughter of Titans, and her sister was Eos, goddess of the dawn. Every civilization throughout history, from Egypt to the Aztecs to the Celts, imagined powerful stories to try and explain the wonder and grandeur that we felt when looking up to the sun and moon.

Looking closely, though, it was the tiny points of light in the night sky that really stretched our imaginings. Some of us thought we were surrounded by a firmament, a vast, complex dome that revolved around us. We strained our weak eyes in the dark to look for patterns and began to fashion great figures in the night sky, from Taurus the bull to Leo the lion, picturing a giant crab and even a scorpion. It made the infinite blackness more familiar somehow, less scary, something we could explain to ourselves. We became fanciful explorers of the heavens, soaring across the arching sky above our heads, finding our way.

No matter how hard we tried, though, we could never have imagined just how complex and wild and vast that sky truly was. When Galileo took our then latest invention, curved, polished glass, and in 1610 built the first telescope, he could suddenly see things that had forever been beyond the capability of our weak human eyes. He was stunned to see that the moon, rather than being the well-known polished, smooth orb he expected, was in fact "uneven, rough, full of cavities and prominences." He was amazed to see tiny planets that were the moons of Jupiter, and strange, inexplicable rings around Saturn. Galileo was like that tottering infant, taking his first steps across the cosmos. With every discovery his imagination soared further, and his writings challenged the fundamental beliefs of many. He had built a window into our solar system, through which we could begin to understand the galaxy that lay beyond it.

With every subsequent improvement in Galileo's telescope, we have been ever more amazed. Our sun, the vastly powerful giver of heat and light, is in fact just one star among hundreds of billions. It's not even that impressive compared to so many others—Betelgeuse, visible to the naked eye in the constellation Orion, is 500 times bigger and 16,000 times as bright. The biggest star we've seen, UY Scuti, is so unimaginably huge that it could contain five billion of our Suns.

We've seen double stars, spinning next to each other like eternal figure skaters. We've witnessed stars explode into a supernova, sending out shockwaves through space like rippling visible sound from a huge firework. We've detected weird and wonderful pulsating stars, or "pulsars," that somehow spin hundreds of times per second. And the vast, omnipresent Milky Way, which the Egyptians imagined as a Nile across the sky, is in fact just so many stars in such close proximity that they blend into the shared glow of the vast swirl of our galaxy itself.

By 1992 our telescopes had gotten so sensitive that we discovered planets orbiting other suns in our galaxy. Since then we've found thousands of these "exoplanets," to the point that now we can conclude that most stars in the universe have planets. And incredibly, our astrophysicists have started to find collapsed stars so dense that even light can't escape their enormous gravity, making them appear as black holes in space. Their gravity is a force of tremendous power, pulling on even distant stars—in fact

there's a super-massive black hole at the center of our galaxy, like a drain at the center of this cosmic swirl we call home.

As an astronaut, I had the chance to take the latest in *Homo sapiens*'s series of first steps when I climbed out of my spaceship and did a spacewalk. As I orbited, my eyes were naturally drawn to Mother Earth, to the colors of the oceans and the familiar shapes of the continents. But it was when I looked around and then up that I was truly awestruck. Suddenly I was enveloped in all dimensions by a bottomless velvety blackness, an eternity so profound and palpable that I felt I could reach out and somehow touch it. The Milky Way was a familiar river flowing across my sight, and every star we've ever seen was there around me. I felt both tiny and huge—minuscule in comparison to the galaxy, and yet immense in my privilege to see it in this new way. It seemed like I was peeking under the edge of what we barely understand. I was gazing out through my visor, on behalf of all who have gone before, into our very future.

We are all explorers, and exploration is simple. All you need to do is look up. The galaxy awaits.

A GREAT

IN

SPIRAL

SPACE

Clouds of Gas
Primordial gas clouds were
created through the cooling
and contraction of gas. The
first stars were created
inside them before the
formation of the disc.

HOW WAS THE MILKY WAY CREATED?

The process that formed the Milky Way, the galaxy that
contains our solar system, started with the creation of
the first stellar clusters from clouds of primordial gases.
The disc was formed afterward and grew in size thanks
to material added from nearby neighboring galaxies.

Growing Disc
The stars are grouped in
clusters that today form
part of the galaxy's halo.
The gradual increase of gas
created a disc that grew,
attracting even more gas,
which eventually formed new
generations of stars.

Globular clusters, groups of stars found in the galactic halo,
formed from high-density clouds of dust and gas. Billions
of years later, the galaxy had gained enough mass so that its
intense gravity caused it to rotate, forming a flat disc. The
next generations of stars were then created, including the
sun. The disc continued growing by incorporating gas from
its own halo as well as those from neighboring galaxies, and
developed several noteworthy structures: the nucleus with its
central bar and the spiral arms. Currently, the Milky Way is
growing in size as it is stealing material from the Magellanic
clouds, two nearby dwarf galaxies.

Old Stars, New Stars
The oldest stars, found in the globular clusters of the halo,
were created some 13 billion years ago through a process that
took approximately 800 million years. The disc contains stars
of various ages depending on whether they are found in the
central bulge, which contains the oldest stars, or in the arms,
where star formation is still occurring.

EVERYTHING FROM GAS
An illustration showing the four most prominent moments in the formation of the Milky Way, from its origin as clouds of gas and dust (left) to the present day as a barred spiral galaxy (below).

Formation of the Nucleus
As the disc contracted, it simultaneously became flat and began rotating faster, forming a dense nucleus at the center.

The Final Structure
The spiral arms were the last structures to be created. They have a higher concentration of interstellar material along with a central bar.

THE STRUCTURE OF THE GALAXY

From above, the Milky Way looks like a barred spiral galaxy, with the solar system located in one of its arms. A frontal view shows the central bulge along with the disc, which makes up the main structure. An enormous halo surrounds it completely.

FRONTAL VIEW

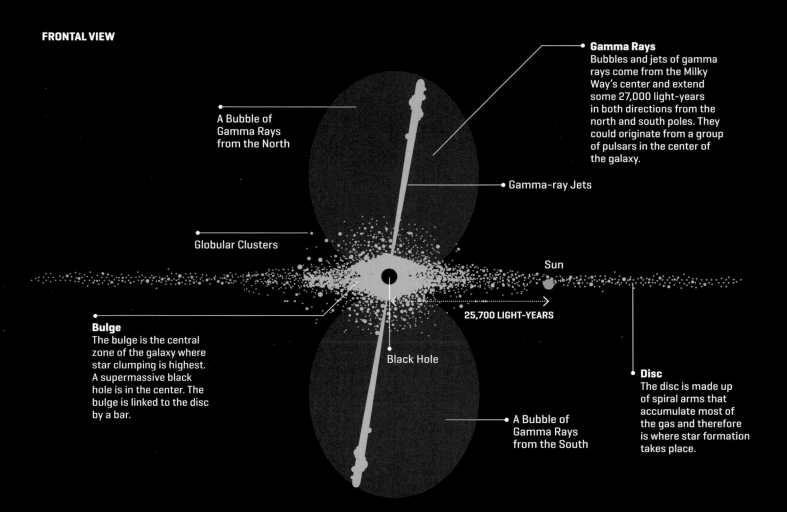

A Bubble of Gamma Rays from the North

Gamma Rays
Bubbles and jets of gamma rays come from the Milky Way's center and extend some 27,000 light-years in both directions from the north and south poles. They could originate from a group of pulsars in the center of the galaxy.

Gamma-ray Jets

Globular Clusters

Sun

25,700 LIGHT-YEARS

Bulge
The bulge is the central zone of the galaxy where star clumping is highest. A supermassive black hole is in the center. The bulge is linked to the disc by a bar.

Black Hole

Disc
The disc is made up of spiral arms that accumulate most of the gas and therefore is where star formation takes place.

A Bubble of Gamma Rays from the South

DISC DIAMETER	Approximately 100,000 light-years
DISC WIDTH	Between 1,000 and 10,000 light-years
TOTAL MASS	Between 0.8 and 2 billion solar masses
AGE OF THE OLDEST STARS	Approximately 13.7 billion years old
NUMBER OF STARS	Between 100 and 400 billion
NUMBER OF PLANETS	Approximately 1 per star
DISTANCE FROM SUN TO CENTER OF GALAXY	Between 24,700 and 26,700 light-years
ORBIT OF SUN AROUND CENTER OF GALAXY	Between 225 and 250 million years

VIEW FROM ABOVE

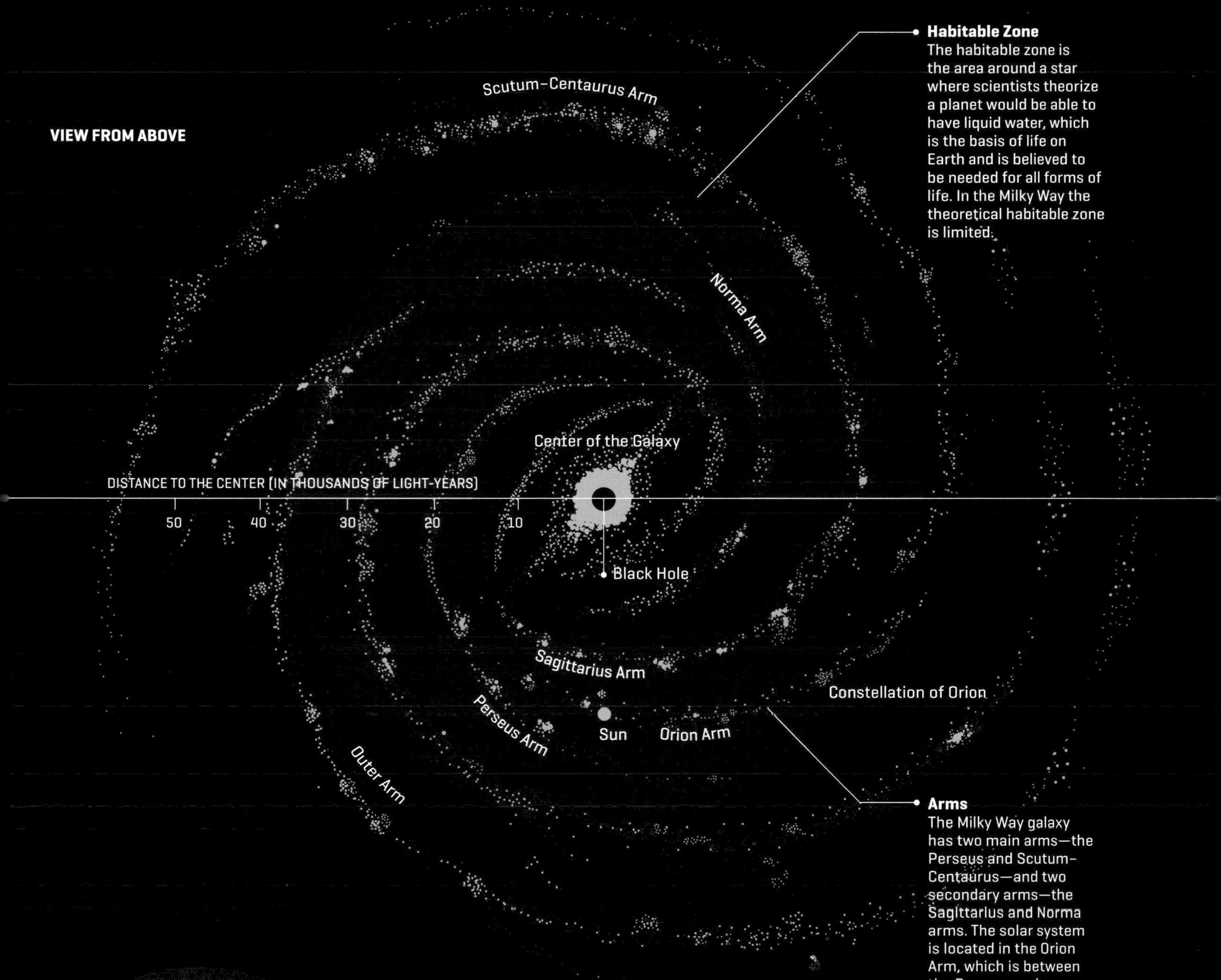

Scutum–Centaurus Arm

Norma Arm

Habitable Zone
The habitable zone is the area around a star where scientists theorize a planet would be able to have liquid water, which is the basis of life on Earth and is believed to be needed for all forms of life. In the Milky Way the theoretical habitable zone is limited.

Center of the Galaxy

DISTANCE TO THE CENTER (IN THOUSANDS OF LIGHT-YEARS)

50   40   30   20   10

Black Hole

Sagittarius Arm

Constellation of Orion

Perseus Arm

Sun   Orion Arm

Outer Arm

Arms
The Milky Way galaxy has two main arms—the Perseus and Scutum–Centaurus—and two secondary arms—the Sagittarius and Norma arms. The solar system is located in the Orion Arm, which is between the Perseus and Sagittarius arms.

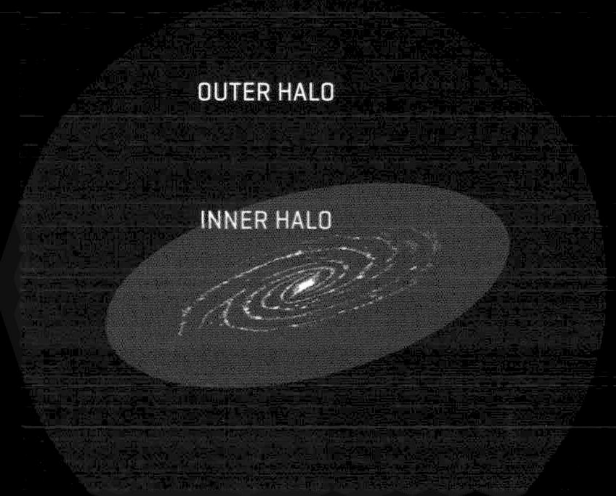

OUTER HALO

INNER HALO

THE GALACTIC HALO
The galactic halo is a structure with a low concentration of gas that surrounds the galaxy. This is where the oldest stars can be found in groups called globular clusters.

TYPES OF GALAXIES

The Milky Way is a barred spiral, one of the five basic classifications of galaxies. This categorization uses only their common morphological, or structural, characteristics without looking at other aspects, such as the rate of star formation and nucleus activity.

Elliptical

1 Elliptical galaxies, such as the M60 galaxy shown here, are characterized as having essentially an ellipsoidal shape without a clearly definable structure. Its stars orbit around a gravitational center in three dimensions. They generally have little interstellar material.

Spiral

2 Spiral galaxies, like NGC 5559, have a well-defined structure around a main plane. The majority of stars orbit in the galaxies' proximity, forming a disc. They are named after the arms that expand from their center and share similar characteristics with lenticular galaxies.

Barred Spiral

3 Many spiral galaxies have a central bar structure that connects the galaxies' arms. Interstellar material is channeled through these structures, from which new stars are created. NGC 1365 is a barred spiral galaxy. The Milky Way also has one of these bars.

Lenticular

4 Lenticular (lens-shaped) galaxies have characteristics of both elliptical and spiral galaxies. They have a disc, like spiral galaxies, but have little interstellar material, like elliptical galaxies. They lack arms, but they can look like they have a spiral structure. An example is NGC 6861.

② ③ ④ ⑤

Irregular

5 Among the most irregular galaxies are some extremely chaotic ones without any sign of a nucleus or spiral structure. They could be spiral or elliptical galaxies before undergoing distortions that occur when they enter the gravitational influence of other galaxies, as in the case of NGC 1427A.

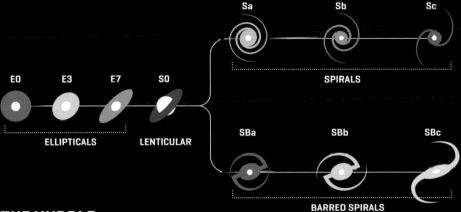

Sa Sb Sc

SPIRALS

E0 E3 E7 S0

ELLIPTICALS LENTICULAR

SBa SBb SBc

BARRED SPIRALS

THE HUBBLE SEQUENCE

The American astronomer Edwin Hubble developed one of the first galaxy classification systems, which is still used today. Hubble defined three basic shape types: elliptical [E], spiral [S], and barred spiral [BS]. Lenticular galaxies, with both spiral and elliptical characteristics, are also included [S0]. The subtypes of elliptical galaxies rely on their appearance: from 0 for an almost circular appearance to 7 for the most flattened shape. Subtypes of spiral galaxies—a, b, c, and so on—indicate how tightly their arms are wound, with subtype a representing a bright nucleus with the most compact arms.

MOVEMENTS IN THE MILKY WAY

Stars, planets, and other celestial bodies are constantly moving around the Milky Way's arms. But that isn't all; the galaxy itself rotates and travels through space at a breathtaking speed.

STELLAR MOTION

Stars in the Milky Way are found in the nucleus, the disc, and the galactic halo. Each group has a specific pattern of movement around the galaxy, as can be seen in this illustration.

Halo
Stars in the halo travel around the galactic disc with random orientations.

Bulge
Stars in the bulge also orbit with random orientations

Disc
All stars in the disc travel in circular, undulating paths in the same direction.

The Milky Way rotates clockwise when one is looking at its northern pole. One of the results of this rotation is the tendency to form spiral structures, which occur when the material farther away from the center has a lower rotational velocity, the speed at which it rotates. The arms move in a wave motion and don't move in relation to the stars; they move independently of one another. Our galaxy moves in relation to other galaxies that are orbiting it at more than one million kilometers (621,371 mi) per hour. It is also moving closer to Andromeda, the nearest major galaxy, with which it will eventually fuse.

The Galactic Neighborhood

Stars in the Milky Way orbit the center of the galaxy, just as the planets and other objects in the solar system orbit the sun. They do not move closer to the galactic nucleus even though it pulls them in with its enormous gravity. Stars also move independently of each other with different velocities, depending on where they are in the galaxy.

A CHANGING STRUCTURE
The structure of the Milky Way's arms has evolved over billions of years. Far from being stable structures, the arms break apart, and new ones are formed over time.

TRAVELING THROUGH THE GALAXY

Just like other stars in the galaxy, the sun orbits around the Milky Way's center, moving north to south and vice versa in relation to the galactic plane, the plane on which most of the galaxy's mass lies. It has a back-and-forth movement caused by the gravitational pull of the nucleus on the one hand and the rotation and attraction of bodies on the other side of the disc on the other hand. The movements of the objects in the solar system also influence our star's orbit. The result is a corkscrew movement that the sun follows as it orbits the galaxy approximately every 235 million years.

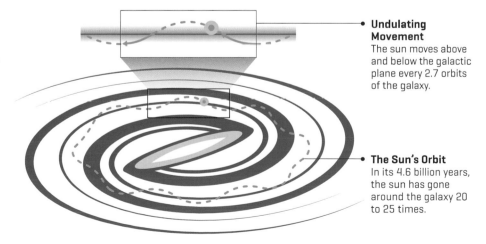

Undulating Movement
The sun moves above and below the galactic plane every 2.7 orbits of the galaxy.

The Sun's Orbit
In its 4.6 billion years, the sun has gone around the galaxy 20 to 25 times.

INTERSTELLAR GAS AND DUST

Observing the Milky Way cross the night sky is a spectacle all in itself, but looking at the insides of this bright zone, we can see even more precious treasures—colored, whimsical clouds of gas and dust.

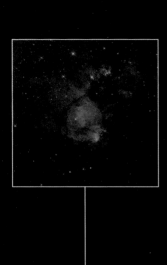

Nebula NGC 896
Some 7,500 light-years away in the Perseus Arm is part of the Heart Nebula. The emissions associated with an open cluster like this are partially obscured by bands of dust.

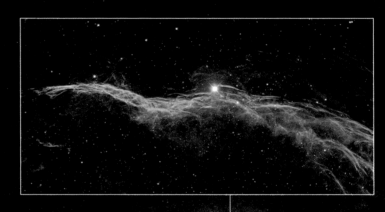

Veil Nebula
This nebula, made from hot, ionized gas and located some 1,500 light-years away, is a remnant of a supernova, an immense explosion of a star that reached the end of its life, that took place several thousand years ago. Its diameter is approximately six times that of Earth's moon.

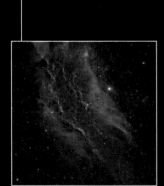

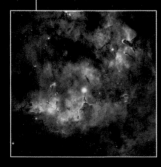

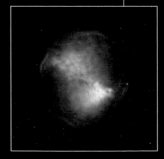

Nebula NGC 7822
Some 3,000 light-years away, this is a region of star formation that includes the Sharpless 171 emission nebula and the young cluster of stars known as Berkeley 59.

California Nebula
Some 1,000 light-years away, this emission nebula has a relatively low brightness, making it easier to study using long-exposure photography.

Dumbbell Nebula
Some 1,200 light-years away, this is a planetary nebula made up of an envelope of bright, ionized gas ejected by a star at the end of its life.

Interstellar nebulae are regions where the density of gas and dust is higher than average. It is believed that they could be the result of an increase in interstellar material, although they could also be the remains of stars that released their material, sometimes violently through cosmic explosions. In the first case, the nebulae could evolve and allow new stars to form.

Lights and Darks

Depending on the way we observe nebulae, they are classified as emission or absorption nebulae. The former are the most common, and they shine due to their gases being excited by ultraviolet radiation from nearby hot stars. The latter do not have nearby stars, so they do not shine and can be seen only in contrast with emission nebulae that are farther away.

Carina Nebula

At a distance between 6,500 and 10,000 light-years away, Carina is a large emission nebula that orbits open clusters. Carina has some of the largest and brightest stars in the Milky Way, including Eta Carinae and HD 93129A, which is actually a system of three stars.

Flame Nebula

This emission nebula, located between 900 and 1,500 light-years away, has a characteristic reddish color due to the excitation of gas from the ultraviolet radiation of Alnitak, a nearby star.

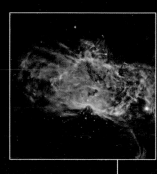

Horsehead Nebula

This nebula, located some 1,500 light-years from Earth, is a cloud of cold gas, which is part of the Orion Complex, made up of both emission and absorption (also called dark) nebulae and young stars. The Horsehead Nebula, seen here in infrared, sits next to the emission nebula IC 434.

Trifid Nebula

An HII region of star formation some 5,500 light-years away, the Trifid Nebula is comprised of an open cluster, an emission nebula, and an absorption nebula, giving it a triangular shape with rounded corners.

THE NEBULAE OF THE MILKY WAY

Interstellar nebulae tend to be concentrated in the discs of spiral galaxies or in any region of irregular galaxies where new stars are formed. Here are representations of the most important nebulae in the Milky Way's disc.

THE GALAXY
IN THE NIGHT SKY

The Milky Way appears as a luminous band that divides the celestial sphere in half along a highly inclined plane with respect to the celestial equator.

There is a great variety of instruments and resources available to identify different astronomical objects. Since ancient times, one of the most used resources has been the constellations, imaginary figures made from combining stars to create different regions in the heavens. The boundaries for most of the 88 constellations today comply with guidelines established by the International Astronomical Union between 1928 and 1930. It is difficult to know the

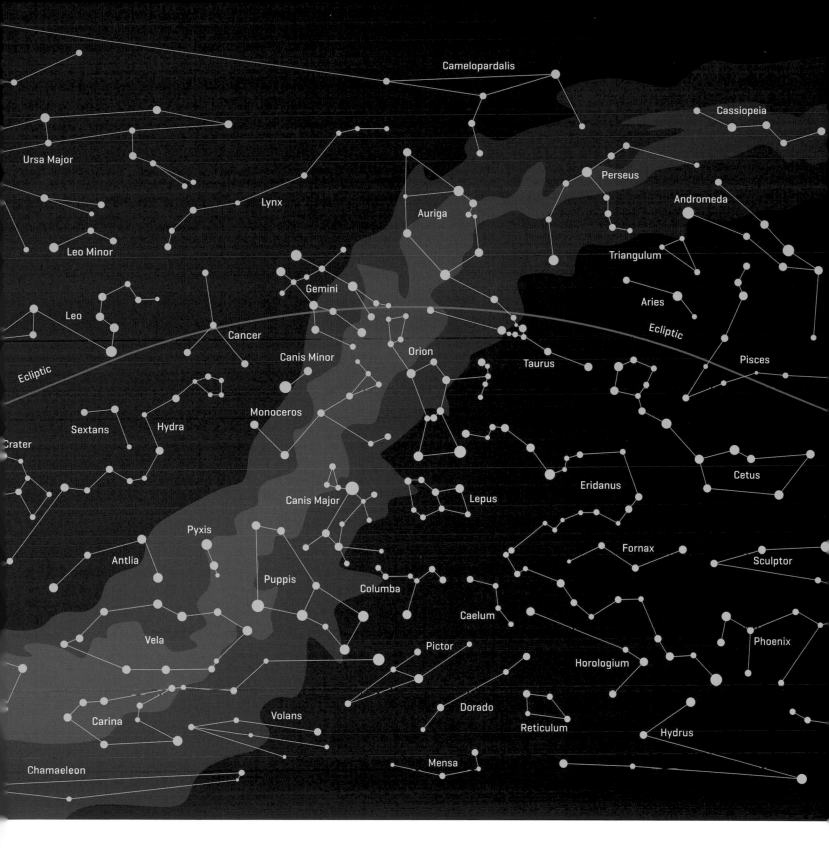

origin of each one exactly, although some refer to Mesopotamian cultures, and half were devised by the Greeks.

Signs of the Zodiac

The most well-known constellations are those of the zodiac, which mark the relative positions of the sun during the year in its journey observed from Earth, following a path known as the ecliptic. Given that the galactic plane is inclined with respect to Earth's plane of orbit, the Milky Way passes only occasionally between the zodiac constellations, between Taurus and Gemini on one side and Scorpio and Sagittarius on the other. It is in this last constellation where we find the galactic center.

THE WHOLE PICTURE

This cylindrical projection of the universe from Earth shows the Milky Way, the constellations, and the ecliptic, a sinusoidal line marking the relative positions of the sun according to Earth's orbit. The two celestial hemispheres are shown in horizontal symmetry with respect to the equatorial plane defined by Earth's rotation.

THE INVISIBLE MILKY WAY

Today we use powerful telescopes that can capture diverse electromagnetic wave frequencies, not just visible light.

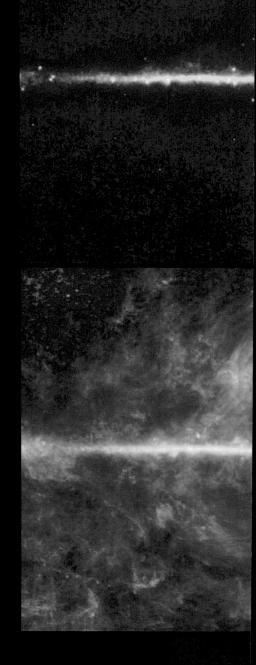

Gamma Rays

1 A map of the galaxy using gamma rays captured by NASA's Fermi Gamma-ray Space Telescope. The telescope has identified all known sources of gamma rays, including some that have been attributed to dark matter.

Near Infrared

2 This image shows the infrared frequencies found just below visible light. In this case, the brightness is not made by dust but by relatively cold giant stars found in the disc and the nucleus.

Microwaves

3 This map, one of the most precise ones made of microwaves, is based on data obtained by the European Space Agency's Planck space observatory. Here you can see the bright band of gas and dust in the galactic plane arching over hundreds of thousands of light-years.

Far Infrared

4 This is made from infrared radiation of the lowest frequencies. At these frequencies, stars emit little radiation, and almost all come from extremely cold clouds of dust. However, they are warm enough to be detected.

Radio Waves

5 Interstellar medium, the matter and radiation that exist between star systems, obscures visible light but does not affect radio waves. This map shows a distribution of cold clouds of neutral hydrogen gas that emit radiation at this frequency.

X-rays

6 Thanks to the x-ray spatial observatories from NASA (Chandra) and the European Space Agency (XMM-Newton), we can obtain information about the Milky Way's center.

THE GALAXY'S EMISSIONS
This image of the Milky Way, created by the European Space Agency's Planck space observatory, is a multifrequency, all-sky image of the microwave sky showing mixtures of gas, charged particles, and various types of dust.

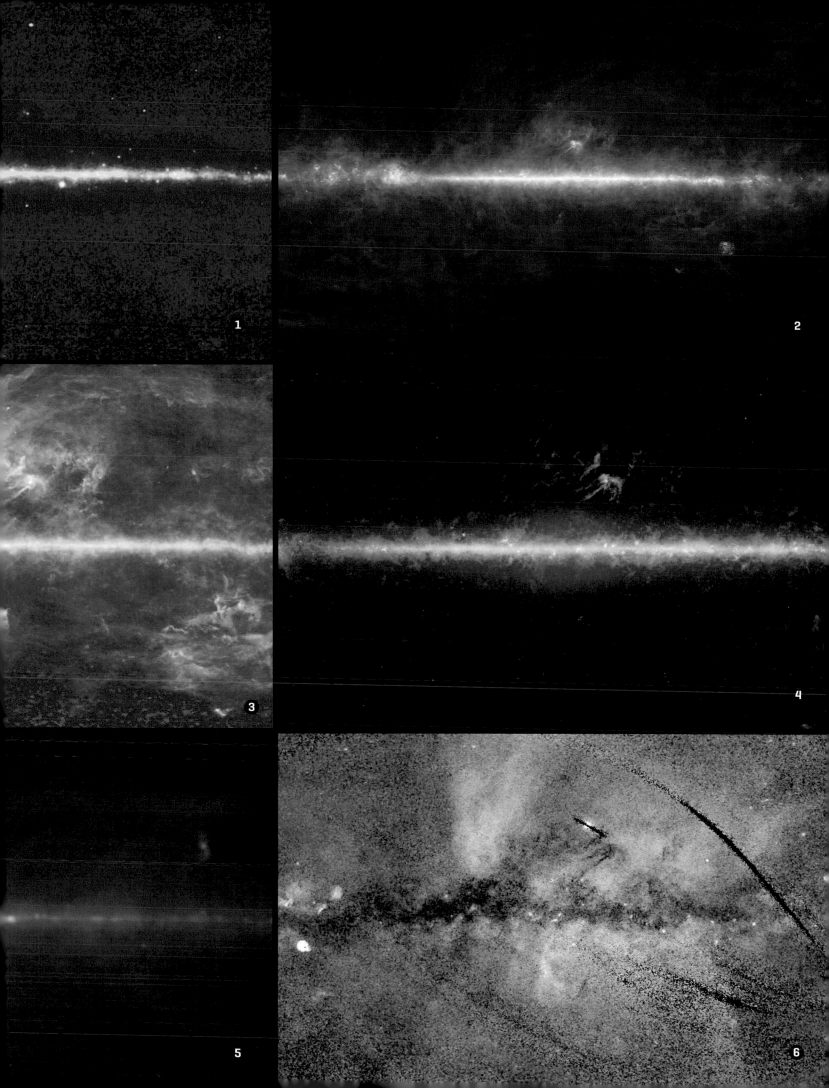

THE PARTS OF
OUR GALAXY

The Milky Way's center, as seen from Earth, contains the highest density of stars in the galaxy, but not the brightest ones, due to the large amount of gas and dust that dims their light. It also contains a supermassive black hole, a region from which nothing, including light, can escape.

THE GALACTIC HALO

Just as with other spiral galaxies, the Milky Way is surrounded by an enormous, spheroidal halo made up of old stars, hot and light gases, and dark matter, which emits no light, can't be directly observed, and is believed to make up 80 percent of the universe.

There is hardly any star formation occurring in the halo that surrounds the Milky Way, as the density of sufficiently cold gases that could collapse to create stars is quite low. The halo's stars are old and tend to be gathered in globular clusters. These stars, which could have been captured from the Milky Way's satellite galaxies, orbit the galaxy in an unconventional way; they can have very inclined, irregular, and even retrograde orbits. It appears that the halo's gases extend for hundreds of thousands of light-years, and its mass could be comparable to the rest of the ordinary matter in the Milky Way. Even so, it is calculated that the total mass of our galaxy is much larger; it is estimated that the amount of dark matter is 5 to 10 times the amount of ordinary matter. This proportion fluctuates based on the distinct estimates that are taken for the mass of the gas.

Rotating Void

The halo's gas, much less dense than any vacuum generated on Earth, can reach 2.5 million degrees and is detectable as x-rays. Recent studies indicate that hot gas in the Milky Way's halo is spinning at a comparable speed and in the same direction as the galaxy's disc, which contains our stars, planets, gas, and dust. Consequently, this rotation will allow more precise models to be made about the formation and development of our galaxy. The halo can be divided into two parts: the interior, which is flat, and the exterior, which is more voluminous. The stars in the halo's interior are younger than those in the exterior and tend to rotate in the same way as the disc; the stars in the exterior rotate the other way and are believed to be the remains of smaller, absorbed galaxies.

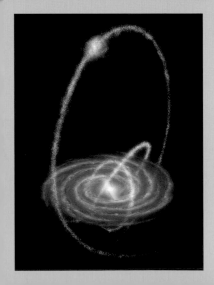

REMNANTS OF GALACTIC COLLISIONS

There are diverse zones in the halo that have a higher stellar density and are thought to be the remnants of galactic collisions. One of these zones is known as the Virgo Stellar Stream. It has an almost perpendicular orientation to the disc and occupies a significant area of the celestial sphere. It is widely believed to be associated with the Sagittarius dwarf spheroidal galaxy, which is being absorbed by the Milky Way.

AN ENORMOUS CLOUD OF GAS
The halo's gases seem to extend much farther than the Magellanic clouds, two characteristic dwarf galaxies in the southern sky. They are located more than 150,000 light-years away and appear to the left of the Milky Way in this artist's depiction.

Two Types of Stars

The Milky Way has two types of stars, known as Population I and Population II. The first type are young stars, wich are rich in heavy elements, and have a short life; the sun is one of these. Population II stars are older, generally have fewer heavy elements, and live for a much longer time; stars in the bulge and galactic halo fall into this category. Stars found in the disc, which are Population I stars, arrange themselves into irregularly shaped open clusters. Stars found in the halo and the bulge form spherical globular clusters and are Population II stars. The globular clusters of the halo are more abundant near the galactic center.

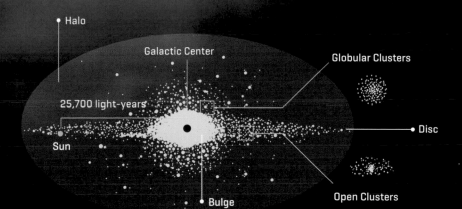

Halo
Galactic Center
Globular Clusters
25,700 light-years
Sun
Disc
Bulge
Open Clusters

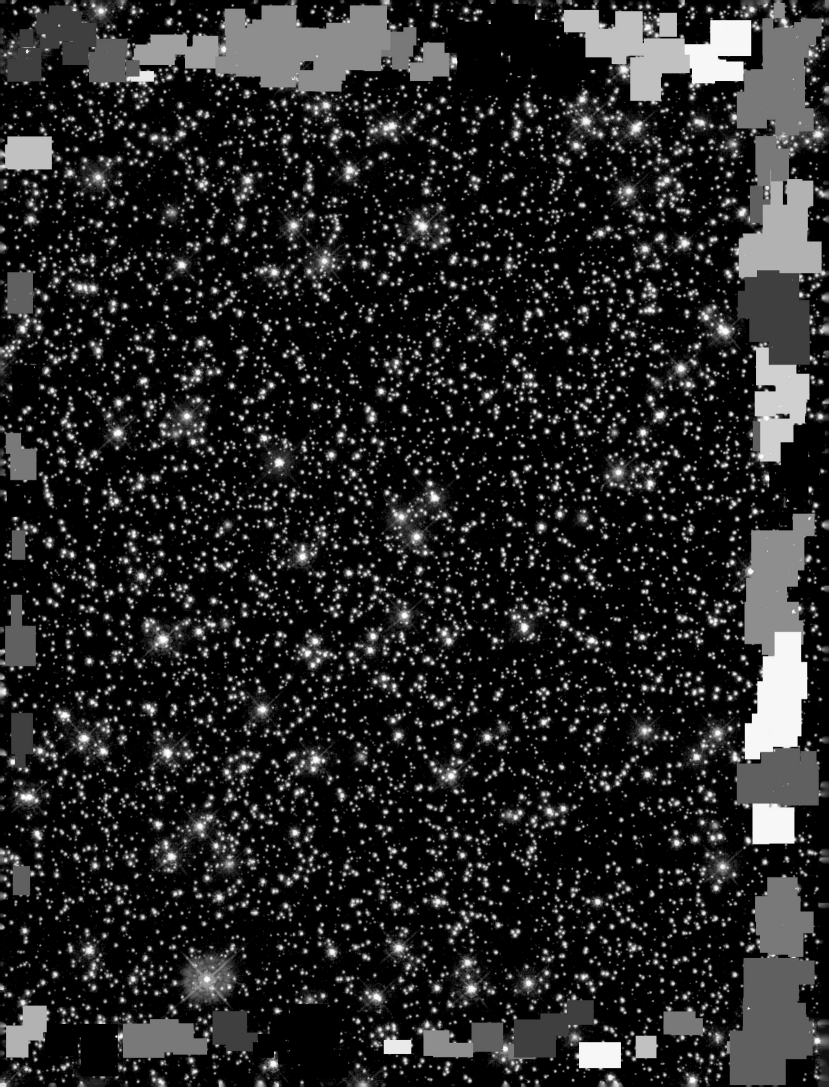

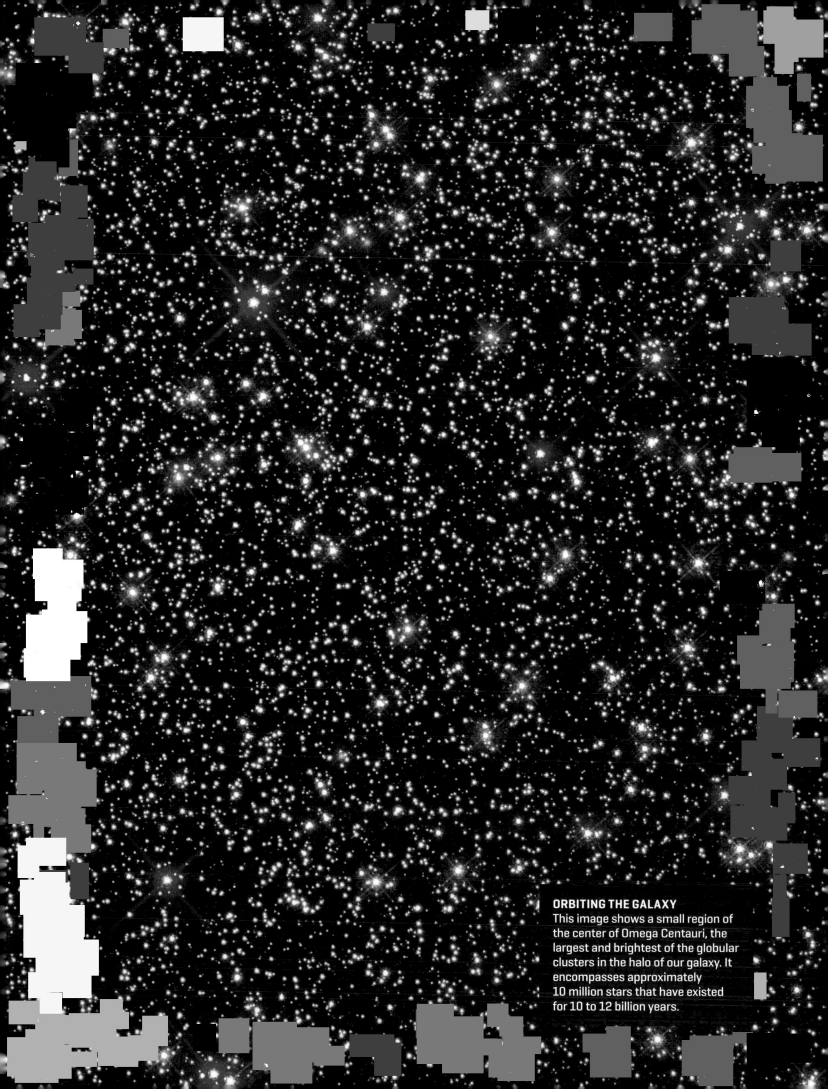

ORBITING THE GALAXY
This image shows a small region of the center of Omega Centauri, the largest and brightest of the globular clusters in the halo of our galaxy. It encompasses approximately 10 million stars that have existed for 10 to 12 billion years.

THE GALAXY'S DISC

The galactic disc is not completely flat but is shaped rather like a warped vinyl record exposed to high temperatures. Although we have known this for decades, it has only been explained satisfactorily within the past few years.

One of the first explanations for the shape of the galactic disc suggested the gravity of nearby dwarf galaxies, known as the Magellanic clouds, was a possible cause. This hypothesis was discarded after a calculation of the joint mass of both proved to be too small to have caused this effect. However, recent investigations now show that the Magellanic clouds are responsible, but the explanation is more complicated. Using computer-generated models, the distortion of our galaxy's disc has been proven to be a result of the Magellanic clouds moving through a halo of dark matter, which would act as an amplifier of their gravitational influence and thus provide the force to warp the galactic disc.

Like a Drum

The model used to simulate the distortion revealed a surprise: this interaction would create a vibration in the Milky Way's disc, similar to the resonance of a drum. According to this hypothesis, the distortion that we can perceive as static would be a snapshot of an instant of a slow, vibratory movement.

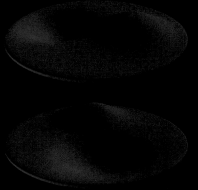

Vibratory Movements
Computer models have shown that the distortion of the disc would correspond with a slow, resonating vibratory movement. The image above illustrates two moments of distortion.

THE HYDROGEN LAYER

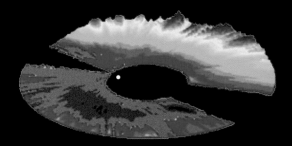

This detailed representation of the Milky Way's distortion has been possible thanks to neutral hydrogen emissions from the gaseous disc, which extend beyond the zones of higher stellar density. The color contours deform "upward" with respect to the galactic plane, while the grays deform "downward." The point situated to the left of the central circle indicates the position of the sun.

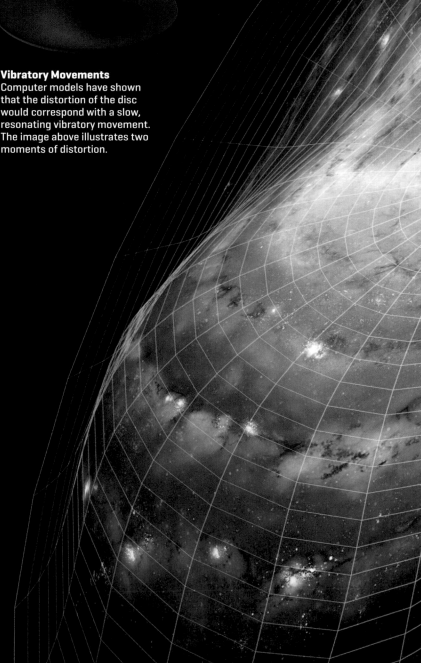

WAVY DISC
The distortion of the Milky Way's disc has been exaggerated in this rendering to show how it folds on both sides of the galactic plane. This distortion is attributed to the Magellanic clouds (represented to the right of the disc), although it is thought to be amplified by the halo of dark matter that envelops the clouds.

THE SPIRAL ARMS

The Milky Way's galactic disc is organized in spiral arms that have a high density of interstellar dust and gas. The concentration and nature of the gas are not always uniform, so it can be split into different regions.

The arms of the galaxy have a high density of interstellar material from which new stars are formed, resulting in a rejuvenated star population. Due to the elevated density, compressed interstellar gas collapses to form molecular clouds and massive stars. The arms contrast with the regions located between them, which accumulate less gas and consequently are incapable of creating massive stars. This results in low brightness and visibility.

Interstellar Material

Stars are made of very hot gas and a small amount of dust particles. The gas is made up of three parts hydrogen to one part helium, along with trace amounts of other elements. Based on their characteristics, four different types of regions can be differentiated: coronal gas, neutral hydrogen (HI) and ionized hydrogen (HII) regions, and molecular clouds.

Density Waves

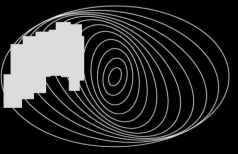

According to the most accepted theory, the spiral arms of galaxies are regions that had a greater density than others at a specific moment, so a wave distortion would propagate these variations in material concentration throughout the galaxy. In other words, the spiral arms are regions in which a larger quantity of material is concentrated at any given time. Stars that occupy these regions don't spin along with the arms; instead, the majority enter and leave the spiral structure of their own accord.

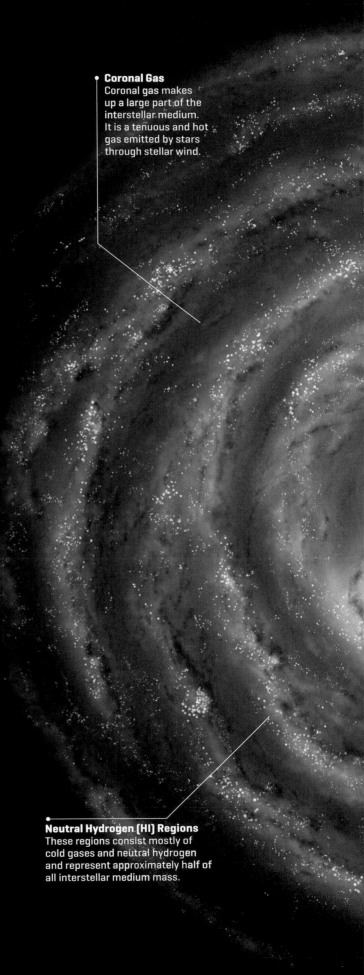

Coronal Gas
Coronal gas makes up a large part of the interstellar medium. It is a tenuous and hot gas emitted by stars through stellar wind.

Neutral Hydrogen (HI) Regions
These regions consist mostly of cold gases and neutral hydrogen and represent approximately half of all interstellar medium mass.

Ionized Hydrogen (HII) Regions
These regions comprise
a small part of the mass of the
interstellar medium but are of
great interest, as they contain
ionized hydrogen, indicating the
presence of recently formed stars.
They can be distinguished by their
characteristic reddish color.

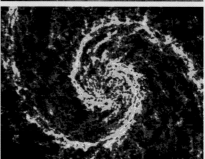

THE WHIRLPOOL GALAXY

The orientation of this spiral galaxy
is similar that of the Milky Way, but
provides us with a view that we could
never see of our own galaxy. The top
image shows the way it can be seen
with visible light; the lower image
shows frequencies characteristic of
carbon monoxide, which is abundant
in molecular clouds. The reddish zones
have a major concentration of gas.

Molecular Clouds
Although relatively small,
they have a high enough
density and a low enough
temperature to allow the
existence of molecular
hydrogen, which is associated
with star formation, along
with carbon monoxide and
other molecules.

Sun
The Orion Arm contains the sun
and the solar system, which
formed some 4.6 billion years
ago as a result of the split of
a giant molecular cloud. Most
likely, the cloud collapsed due
to shock waves arising from a
nearby supernova.

STELLAR NURSERIES
The Milky Way's arms, just as
in any spiral galaxy, are the
regions in which the rate of

A BARRED SPIRAL

The galactic nucleus is oriented in the direction of the constellation Sagittarius, where the Milky Way can be seen at its brightest. Although it is hidden by large amounts of interstellar dust, it is possible to see that it has a barred structure.

The nucleus has the highest concentration of stars in the Milky Way, with a density that is only surpassed in our galaxy by some globular clusters. Its stellar mass is estimated to be around 20 billion times that of our own star, the sun, and its luminosity more than five billion times that of the sun. In the 1990s, scientists began to suspect that the galaxy had a barred structure; since then the body of evidence points in this direction. The exact form of the bar continues to be a subject of discussion, but it can reach a longitude of some 30,000 light-years and an inclination of some 45 degrees with respect to the imaginary line that connects us to the galactic center.

Observing the Nucleus

Interstellar dust found between Earth and the galactic nucleus allows neither visible nor ultraviolet radiation nor low-energy x-rays to pass. In order to gain more information about the nucleus, we have to use radiation of higher and lower frequencies, in other words, high-energy x-rays and gamma rays, as well as infrared and radio waves.

Nature of the Bar
The central bars of galaxies can appear to be solid structures, but they are actually made up of density waves. The gas near the nucleus moves faster than gas farther away, contributing to the creation of the bar.

THE STARS OF THE CENTRAL BAR

The stars in the central bar can be identified by their velocities, moving closer or farther away from Earth according to which side of the galactic center we see them from. In this map from the Sloan Digital Sky Survey project, the circles indicate regions of explored space. Those marked with "X" show places where stars have been detected that move away from us, while on the other side, the circles with dots indicate places where they are coming toward Earth.

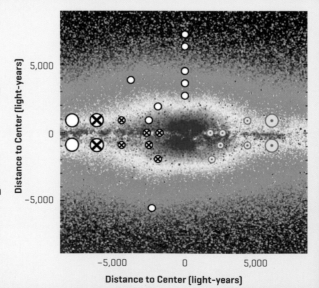

Distance to Center [light-years]

Distance to Center [light-years]

THE GALAXY'S BAR
This artist's rendering depicts how the barred structure in the nucleus of the Milky Way would be appear from outside the disc.

OTHER BARRED GALAXIES

NGC 1300
Approximately 60 million light-years away from Earth, NGC 1300 has a smaller spiral structure in the middle of its nucleus.

NGC 1672
This galaxy has asymmetric spiral arms with large differences in brightness. It is located some 60 million light-years away.

NGC 7479
Located some 105 million light-years away, NGC 7479 is known for its asymmetric structure as well as for its brightness and long central bar with a high activity of star formation.

THE CENTER OF THE MILKY WAY

The central region of the Milky Way is one of the most complex and turbulent areas of the galaxy. Although difficult to observe, the stellar density and the existence of a supermassive black hole have been determined.

Less than four light-years from the galactic center, the central region of the Milky Way is an area with a high density of stars. Although many of the stars have a similar mass to the sun's and are older, others, known as S stars, which are more massive and seem to have formed only a few tens of millions of years ago, have been identified. The central cluster is the largest and densest in our galaxy. Its stellar concentration is the equivalent of packing in a million stars between the sun and its closest star, Proxima Centauri. This concentration and high velocities of stars found here cause frequent collisions in the central cluster.

A Supermassive Center
Stars that are very close to the galactic center move faster than those that are farther away, suggesting there is a black hole in the Milky Way's center. Further proof of the existence of this are the jets of gamma rays that appear to come from the center of the galaxy and could be the remains of the violent past of a black hole.

ANTIMATTER IN THE CENTER OF THE GALAXY

Large quantities of positrons (particles that have the same characteristics as electrons but with a positive charge) have been detected around the galactic center. Pairs of these particles and antiparticles cancel each other out, emitting gamma rays that can be used to determine their presence. Emissions near the galactic center (the brightest part) have also been detected above the galactic plane (horizontal structure).

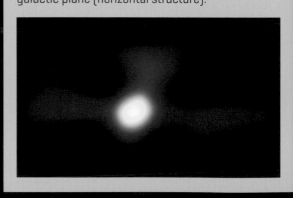

The Heart of the Milky Way's Structure

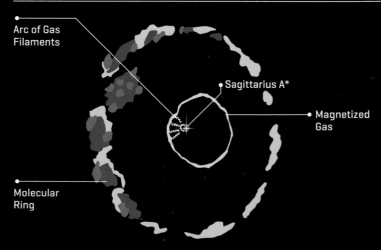

Arc of Gas Filaments

Sagittarius A*

Magnetized Gas

Molecular Ring

The galactic center is surrounded by the molecular ring, which is a group of cold hydrogen clouds, molecular clouds, and nebulae. Inside, there is a vast region of magnetized gas that is connected by filaments of gas to Sagittarius A*, the source of radio waves, and to the exact center of the galaxy.

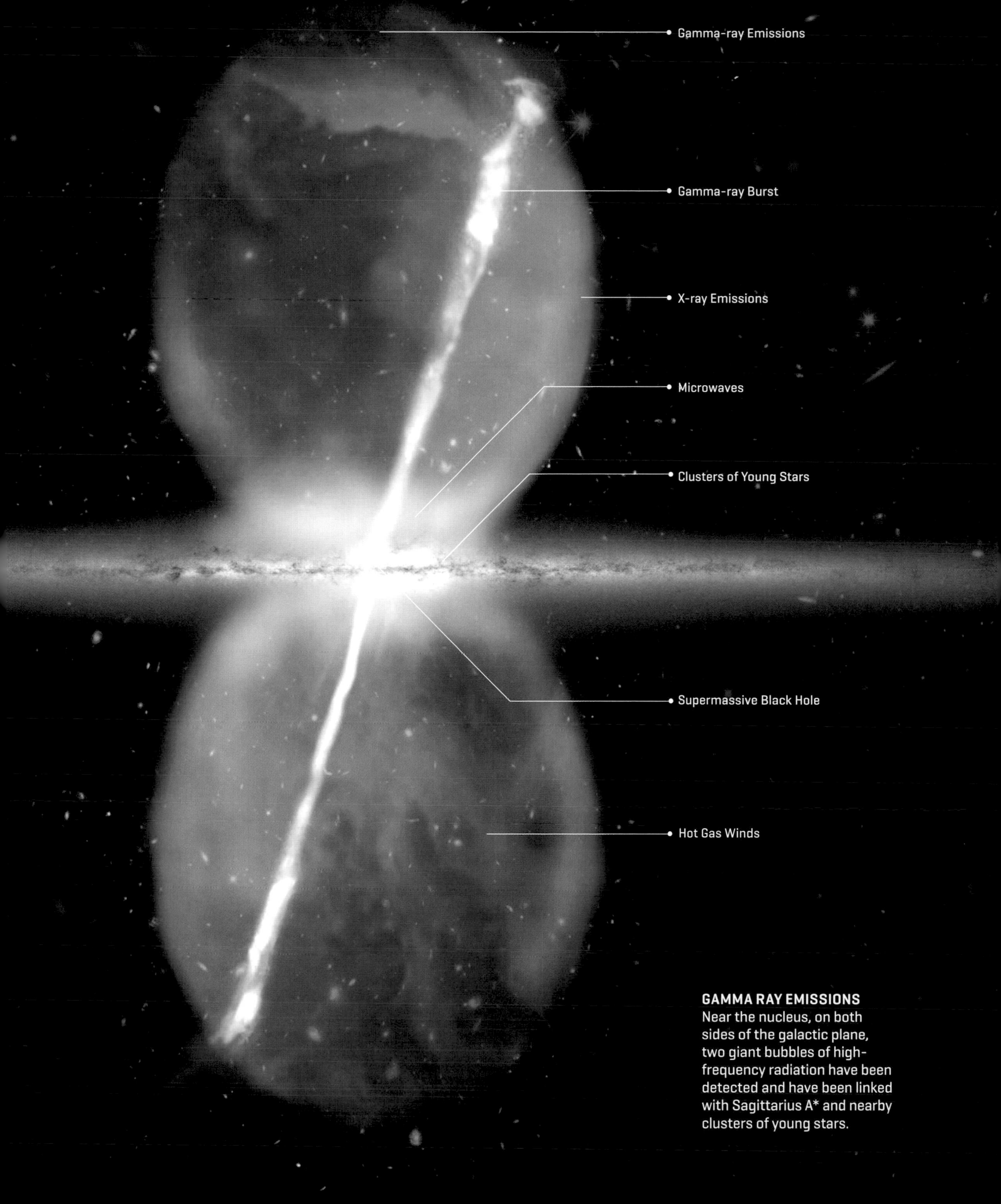

Gamma-ray Emissions

Gamma-ray Burst

X-ray Emissions

Microwaves

Clusters of Young Stars

Supermassive Black Hole

Hot Gas Winds

GAMMA RAY EMISSIONS
Near the nucleus, on both sides of the galactic plane, two giant bubbles of high-frequency radiation have been detected and have been linked with Sagittarius A* and nearby clusters of young stars.

SAGITTARIUS A*

The Milky Way's black hole makes itself known through its gravitational effect, which it exerts on nearby stars, and through radio wave emissions when it manages to consume some interstellar gas and dust. Otherwise, it appears to have consumed almost all of the material that surrounded it.

The exact position of the Milky Way's black hole is marked as Sagittarius A*, a source of radio waves. Although the radiation coming from Sagittarius A* is intense, it is relatively weak for a supermassive black hole. This could explain the lack of gas and dust in its vicinity, causing an extremely low accretion rate, or accumulation of particles, in comparison with active nuclei of other galaxies. It is likely that the Milky Way's black hole formed in the first stages of the galaxy's formation from a huge gas cloud that became increasingly dense over time.

Massive Stars, Subdued

Young S stars, whose mass is higher than the rest of the stars in the central region of the Milky Way, orbit at a higher velocity near the galactic center: some 1,000 kilometers (621.37 mi) per second. This indicates the presence of a black hole. According to this stellar motion, its mass is estimated to be four million times the sun's, confined to a space smaller than Earth's orbit. The enormous gravitational field associated with the black hole rules out the possibility that S stars were formed there; the most probable explanation is that they were captured by it.

An Enormous Center of Gravity
S stars allow for the calculation of the mass of the supermassive black hole in the center because they stand out, and their trajectories can be traced. One of the closest is S2 (red in the diagram), whose orbital period is approximately 15.5 years.

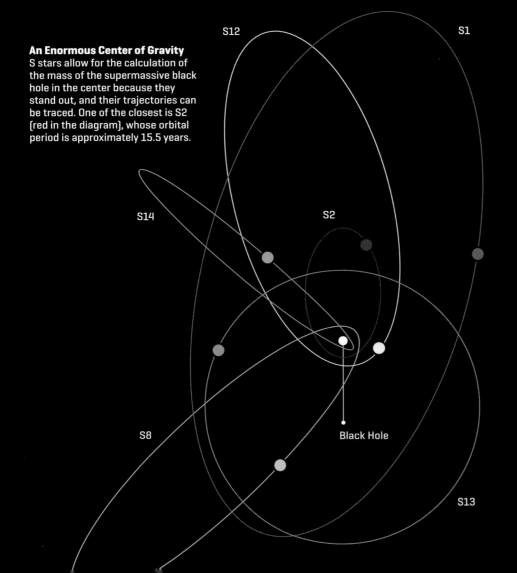

S12

S1

S14

S2

S8

Black Hole

S13

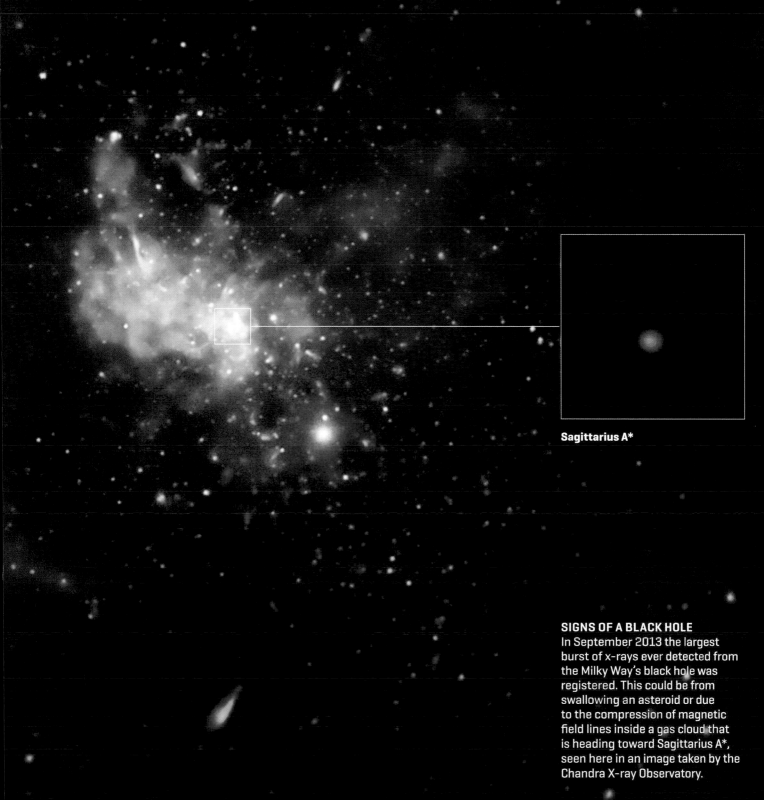

Sagittarius A*

SIGNS OF A BLACK HOLE
In September 2013 the largest burst of x-rays ever detected from the Milky Way's black hole was registered. This could be from swallowing an asteroid or due to the compression of magnetic field lines inside a gas cloud that is heading toward Sagittarius A*, seen here in an image taken by the Chandra X-ray Observatory.

THE CENTER OF THE GALAXY EXPOSED
The clouds of dust that are found around the galaxy obscure visible light but allow infrared light through, revealing its center in a bright yellow color. This is a very dense and compact region of stars. These stars and the gas that is found near the center orbit very quickly. This is where the supermassive black hole Sagittarius A* is located.

WHERE

LIVE

STARS

AND DIE

THE STARS, SOURCES OF MATTER

The large majority of elements on the periodic table, including those necessary for life as we know it, were created thanks to chemical reactions in the cores of stars, then ejected during the violent deaths of the most massive stars.

The Material of a Primordial Universe

1 In the beginning, before the formation of the first stars, only hydrogen, helium, and a small amount of lithium existed in the primordial universe. These were the ingredients that shaped the creation of other elements.

H																	He
Li	Be											B	C	N	O	F	Ne
Na	Mg											Al	Si	P	S	Cl	Ar
K	Ca	Sc	Ti	V	Cr	Mn	Fe	Co	Ni	Cu	Zn	Ga	Ge	As	Se	Br	Kr
Rb	Sr	Y	Zr	Nb	Mo	Tc	Ru	Rh	Pd	Ag	Cd	In	Sn	Sb	Te	I	Xe
Cs	Ba		Hf	Ta	W	Re	Os	Ir	Pt	Au	Hg	Tl	Pb	Bi	Po	At	Rn
Fr	Ra		Rf	Db	Sg	Bh	Hs	Mt	Ds	Rg	Cn	Nh	Fl	Mc	Lv	Ts	Og

	La	Ce	Pr	Nd	Pm	Sm	Eu	Gd	Tb	Dy	Ho	Er	Tm	Yb	Lu
	Ac	Th	Pa	U	Np	Pu	Am	Cm	Bk	Cf	Es	Fm	Md	No	Lr

☐ Necessary Elements for Life

The Elements of the First Stars

2 Many of the essential elements for life, such as carbon and oxygen, were created in the cores of the first stars, and they continue to form in stars like our sun and those with a higher mass.

H								
Li	Be							
Na	Mg							
K	Ca	Sc	Ti	V	Cr	Mn	Fe
Rb	Sr	Y	Zr	Nb	Mo	Tc	Ru
Cs	Ba		Hf	Ta	W	Re	Os
Fr	Ra		Rf	Db	Sg	Bh	Hs

	La	Ce	Pr	Nd	Pm	Sm
	Ac	Th	Pa	U	Np	Pu

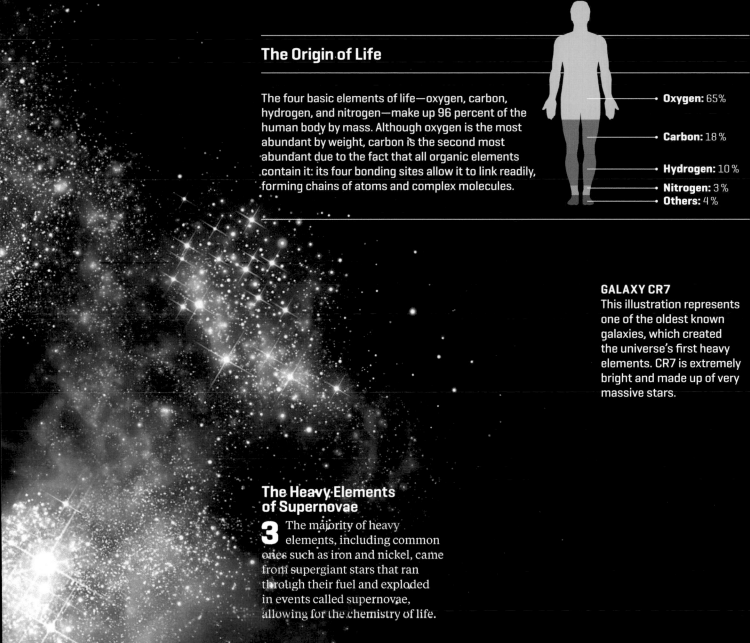

The Origin of Life

The four basic elements of life—oxygen, carbon, hydrogen, and nitrogen—make up 96 percent of the human body by mass. Although oxygen is the most abundant by weight, carbon is the second most abundant due to the fact that all organic elements contain it: its four bonding sites allow it to link readily, forming chains of atoms and complex molecules.

- **Oxygen:** 65%
- **Carbon:** 18%
- **Hydrogen:** 10%
- **Nitrogen:** 3%
- **Others:** 4%

GALAXY CR7
This illustration represents one of the oldest known galaxies, which created the universe's first heavy elements. CR7 is extremely bright and made up of very massive stars.

The Heavy Elements of Supernovae

3 The majority of heavy elements, including common ones such as iron and nickel, came from supergiant stars that ran through their fuel and exploded in events called supernovae, allowing for the chemistry of life.

STELLAR NURSERIES

Our galaxy continues to generate a multitude of stars, using large amounts of interstellar material in the form of dense clouds of gas as raw material.

In star-forming nebulae, the most abundant gas is hydrogen, found in diatomic form (hydrogen dioxide), which creates what is known as molecular clouds. Thousands of these clouds, with a mass at least 100,000 times that of the sun, are thought to populate the Milky Way. They are diffuse enough that they do not contract unless something triggers an increase in density: some kind of gravitational disturbance that gives rise to spinning clumps of gas and dust. When these clumps gain enough mass, they collapse under the compressive force of gravity and begin to form a protostar.

Cold or Hot
The type of stars a molecular cloud creates depends on its characteristics. Colder clouds tend to create less massive stars, while large, hot clouds produce a wider variety, including stars with a higher mass.

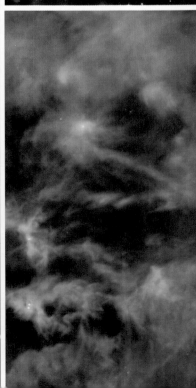

CEPHEUS B
This image made from x-rays and infrared shows the Cepheus B molecular cloud, located 2,400 light-years from Earth. (A light-year is the distance that light travels in one year. One light-year equals 9.46 trillion kilometers, or 5.88 trillion miles.) The presence of protoplanetary discs—rotating discs of dust and gas that surround the core of the developing solar system—in and around the cloud indicate how young its stars are.

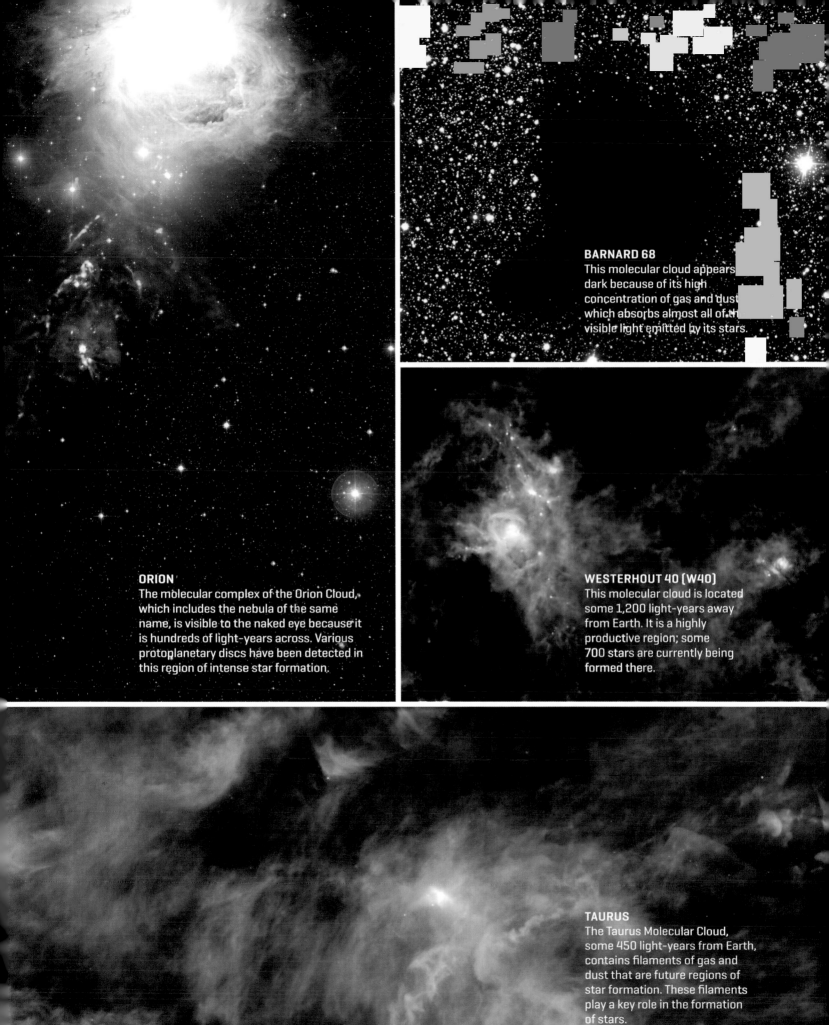

BARNARD 68
This molecular cloud appears dark because of its high concentration of gas and dust, which absorbs almost all of the visible light emitted by its stars.

ORION
The molecular complex of the Orion Cloud, which includes the nebula of the same name, is visible to the naked eye because it is hundreds of light-years across. Various protoplanetary discs have been detected in this region of intense star formation.

WESTERHOUT 40 (W40)
This molecular cloud is located some 1,200 light-years away from Earth. It is a highly productive region; some 700 stars are currently being formed there.

TAURUS
The Taurus Molecular Cloud, some 450 light-years from Earth, contains filaments of gas and dust that are future regions of star formation. These filaments play a key role in the formation of stars.

A COSMIC ROSE
The Rosette Nebula, imaged in infrared by NASA's Spitzer Space Telescope, is a large cloud of gas and plasma in which many massive stars are being formed. This delicate rose-shaped cloud is hiding super-hot stars whose radiation and stellar winds have cleared away layers of dust (shown in green) and gas, revealing a core of colder dust (shown in red).

STARS IN OUR GALAXY

The Milky Way produces fewer stars than it did in the past, so to understand how our galaxy formed we can't just study star formation close to home. Therefore we look to other galaxies with similar characteristics at different cosmic times.

Galaxies like ours have a lot of mass. So when a smaller galaxy gets close enough, the larger one will start to tear it apart and absorb it. For example, the Milky Way is absorbing the Canis Major Dwarf and Sagittarius Dwarf elliptical galaxies and siphoning off material from the Magellanic Clouds. It seems that such mergers happened more often in our galaxy's younger years, until it grew large enough and rotated so rapidly that it flattened into a disc where the next generation of stars could form, including our sun.

Active Despite Its Age

Though the Milky Way is an old galaxy—some 13.5 billion years—it still produces a lot of stars. In fact, new clouds of young stars have been discovered in just the past few years. Massive clusters, which are made up of stars whose total mass is greater than 10,000 solar masses, show the galaxy's level of stellar activity. It is believed that the Milky Way contains around 100 of these massive clusters—a number comparable to other, similar galaxies. Our galaxy creates an average of five to seven new stars each year.

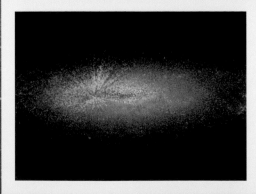

THE AGES OF STARS

This image shows the distribution of red giants in the Milky Way, up to a range of approximately 50,000 light-years away from the galaxy's center. The oldest, some 12 billion years old, are shown in red, the youngest are shown in blue, and the green ones are somewhere in between. This distribution, with the oldest stars in the center and the youngest ones in the disc, confirm that the galaxy's core was created first, then expanded outward.

THE MILKY WAY'S MOST FRUITFUL TIME
This image is a re-creation of the Milky Way as seen from a hypothetical planet 10 billion years ago, during its fastest rate of star creation, some 30 times greater than it is today.

Galactic Layers

| 11.3 billion years ago | 10.9 billion years ago | 10.3 billion years ago |
| 8.9 billion years ago | 6.1 billion years ago | 3.1 billion years ago |

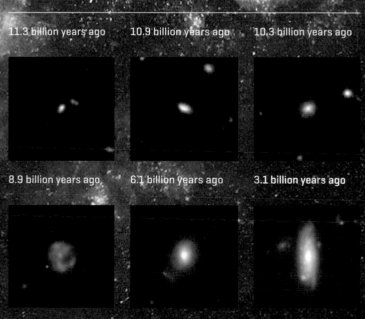

We can better understand the evolution of our own galaxy by observing others at different cosmic distances and times. The youngest (upper row) can be identified by their significant star formation activity; the more mature (lower row) by their spiral structure, which becomes more developed as they evolve.

A DYNAMIC MAP
The H-R diagram is far from static. As a star ages and its physical properties evolve, it will appear on different points of the diagram. Stars spend most of their lives in the main sequence: the cooler and smaller they are, the longer they live there.

M

A

I

N

CLASSIFICATION OF STARS

Stars are arranged in the Hertzsprung-Russell (H-R) diagram according to their luminosity and surface temperature, creating a graphic representation of where stars are in their life cycle.

The H-R diagram, named for astronomers Ejnar Hertzsprung and Henry Norris Russell, is a visual plot of stellar evolution that shows the relationship between a star's temperature (determined by its color, or a spectral analysis of its light) and luminosity (determined by its brightness, or the amount of energy or light it emits). The majority of stars follow a clear pattern: the more luminous the star, the hotter it is. Most stars run along a diagonal band called the main sequence, which is where they spend the majority of their lives. Above the main sequence are the very bright stars: giants and supergiants, the colossuses of the sky. Below it are very faint stars, white dwarfs.

White Dwarfs
This zone is occupied by what were once low- to medium-sized stars similar in mass to the sun. When a star runs out of fuel and sheds its outer layers, its carbon core cools and collapses into a white dwarf, a dense stellar corpse that will continue to cool and fade for billions of years.

Temperature (Kelvin [K])   30,000   10,000

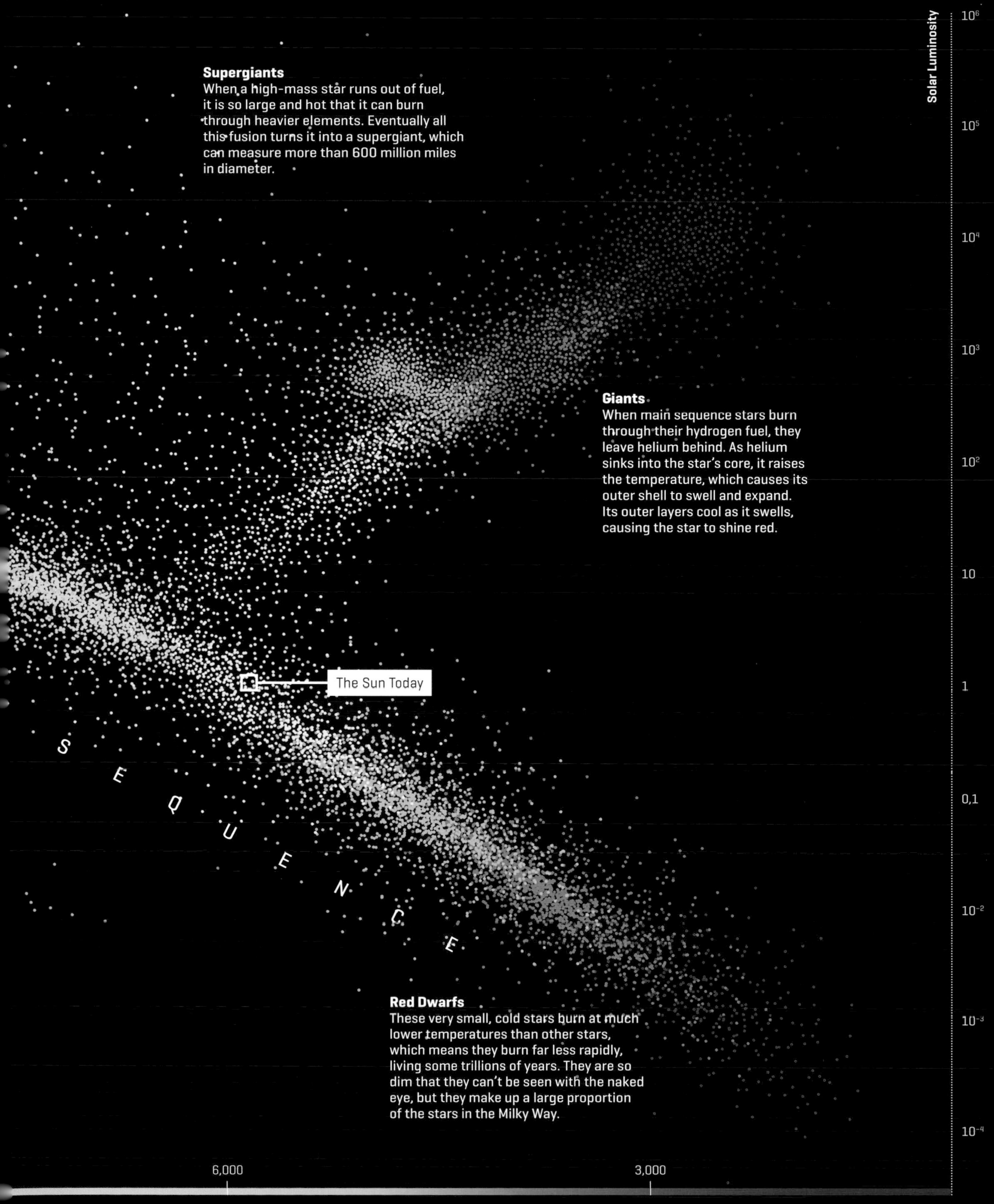

Solar Luminosity

10^6

10^5

10^4

10^3

10^2

10

1

0,1

10^{-2}

10^{-3}

10^{-4}

Supergiants
When a high-mass star runs out of fuel, it is so large and hot that it can burn through heavier elements. Eventually all this fusion turns it into a supergiant, which can measure more than 600 million miles in diameter.

Giants
When main sequence stars burn through their hydrogen fuel, they leave helium behind. As helium sinks into the star's core, it raises the temperature, which causes its outer shell to swell and expand. Its outer layers cool as it swells, causing the star to shine red.

The Sun Today

S E Q U E N C E

Red Dwarfs
These very small, cold stars burn at much lower temperatures than other stars, which means they burn far less rapidly, living some trillions of years. They are so dim that they can't be seen with the naked eye, but they make up a large proportion of the stars in the Milky Way.

6,000

3,000

TEMPERATURE AND SPECTRAL TYPE

Based on their temperatures, which are measured using their emitted electromagnetic spectrum, astronomers classify stars by spectral types, which are directly related to their color and surface brightness.

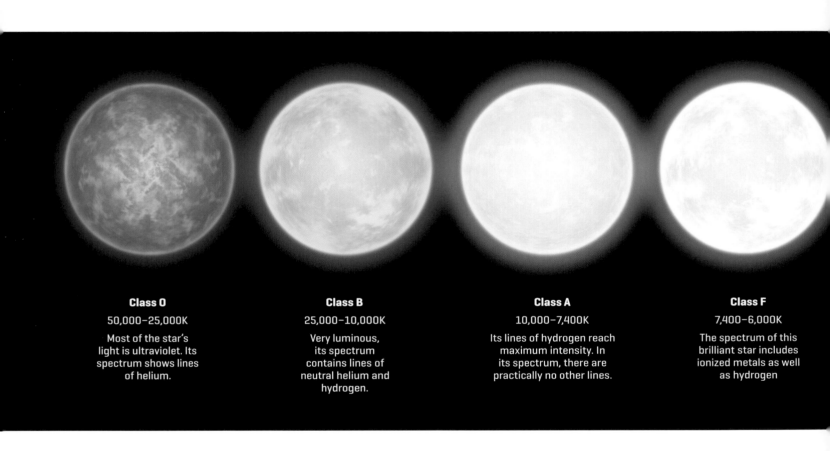

Class O

50,000–25,000K

Most of the star's light is ultraviolet. Its spectrum shows lines of helium.

Class B

25,000–10,000K

Very luminous, its spectrum contains lines of neutral helium and hydrogen.

Class A

10,000–7,400K

Its lines of hydrogen reach maximum intensity. In its spectrum, there are practically no other lines.

Class F

7,400–6,000K

The spectrum of this brilliant star includes ionized metals as well as hydrogen

THE ELECTROMAGNETIC SPECTRUM

The electromagnetic spectrum is the entire existing range of light. Light is electromagnetic radiation: a series of waves characterized by their frequency (how many waves pass by a certain point every second) and wavelength (the distance between the waves' peaks). The larger the frequency, the smaller the wavelength. Gamma rays and x-rays have high-energy, shorter wavelengths, while radio waves and microwaves have low-energy, larger ones. Our eyes can perceive radiation between approximately 390 and 700 nanometers (nm), a range that we call visible light—but in fact, most light is invisible to us. Beyond red is infrared, and beyond violet is ultraviolet.

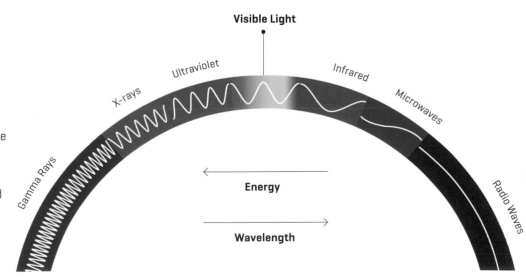

Visible Light

X-rays
Ultraviolet
Infrared
Microwaves
Gamma Rays
Radio Waves

Energy

Wavelength

At the beginning of the 20th century, a group of experts led by American astronomers Edward Pickering and Annie Jump Cannon devoted themselves to analyzing 250,000 photographs of stars, creating a classification system based on their electromagnetic radiation spectrum—the distribution of electromagnetic waves that stars emit and absorb. Even though they observed atmospheres with different chemical compositions, the temperatures of the stars turned out to be a basic factor that could be used to differentiate them.

The Harvard Classification

Pickering and Cannon labeled the stars based on their surface temperatures using the letters O, B, A, F, G, K, and M, with O being the hottest stars and M the coolest (the sun is a G star). This system, called the Harvard classification, is still used today. In recent years other types have been added, such as L and T for substellar (smaller than average) objects known as brown dwarfs.

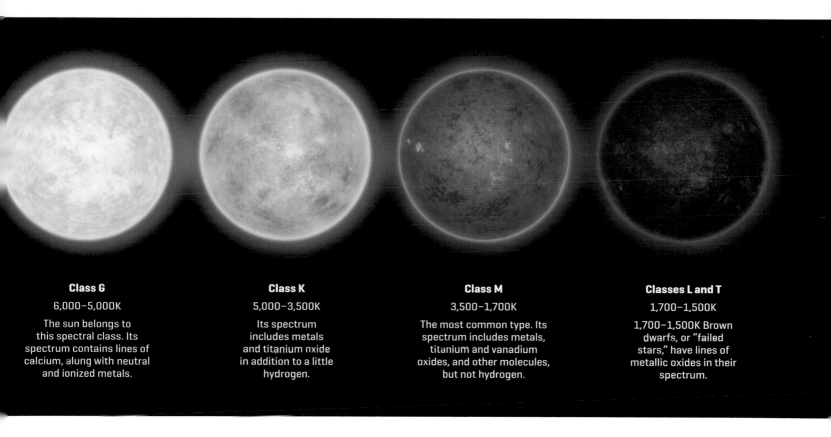

Class G	Class K	Class M	Classes L and T
6,000–5,000K	5,000–3,500K	3,500–1,700K	1,700–1,500K
The sun belongs to this spectral class. Its spectrum contains lines of calcium, along with neutral and ionized metals.	Its spectrum includes metals and titanium oxide in addition to a little hydrogen.	The most common type. Its spectrum includes metals, titanium and vanadium oxides, and other molecules, but not hydrogen.	1,700–1,500K Brown dwarfs, or "failed stars," have lines of metallic oxides in their spectrum.

WAVELENGTH AS AN INDICATION OF TEMPERATURE

The spectral distribution of a star contains information about its surface temperature: very hot stars emit higher-intensity radiation at very short wavelengths, while cooler ones emit it at longer wavelengths. So the hotter the star, the shorter the wavelength of light it emits and the bluer it appears, while the cooler the star, the longer the wavelength of light it emits and the redder it appears.

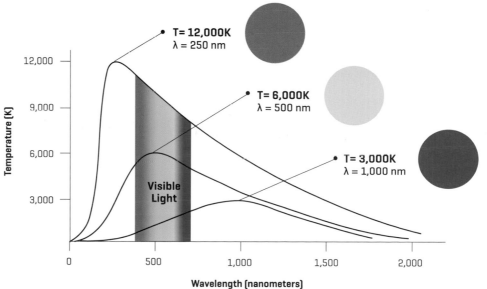

T= 12,000K
λ = 250 nm

T= 6,000K
λ = 500 nm

T= 3,000K
λ = 1,000 nm

Temperature [K]

12,000

9,000

6,000

3,000

Visible Light

0   500   1,000   1,500   2,000

Wavelength (nanometers)

BRILLIANCE AND DISTANCE

The brilliance of a star in the night sky depends not only on its intrinsic luminosity or radiation potential, but also on its distance from Earth.

Using a star's brilliance for comparison has drawbacks, as brilliance depends not only on the strength of the radiation it emits (its luminosity) but also on how far away it is from the point of measurement.

The Apparent Magnitude
Modern astronomers classify brilliance according to a star's apparent magnitude, or its brightness as seen from Earth. The lower the number on the magnitude scale, the brighter a star appears, with 6 corresponding to stars on the edge of naked eye visibility. At every decrease in apparent magnitude (say from magnitude 2 to 1), the star shines 2.5 times more brightly, with negative values representing the brightest objects.

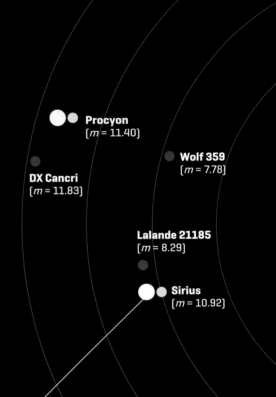

GJ 1061
[m^* = 11.99]

Epsilon Eridani
[m = 10.52]

Procyon
[m = 11.40]

Wolf 359
[m = 7.78]

DX Cancri
[m = 11.83]

Lalande 21185
[m = 8.29]

Sirius
[m = 10.92]

Ross 128
[10.92 light-years]

Sirius
This system is made up of two stars: Sirius A, which is the bright star in Earth's night sky, and Sirius B, a white dwarf.

THE CLOSEST STARS
The distance between us and the stars is so vast that even those nearest to Earth are several light-years away.

* m is apparent magnitude

WHAT IS A LIGHT-YEAR?

Sun	Earth	Jupiter	Interior edge of the Kuiper belt	Exterior edge of the Kuiper belt
	8.31 light-minutes	43.25 light-minutes	4 light-hours	7 light-hours
	150 million km	778 million km	4.317 billion km	6.476 billion km [4.024 billion m
	[93.2 million mi]	[483.4 million mi]	[2.682 billion mi]	

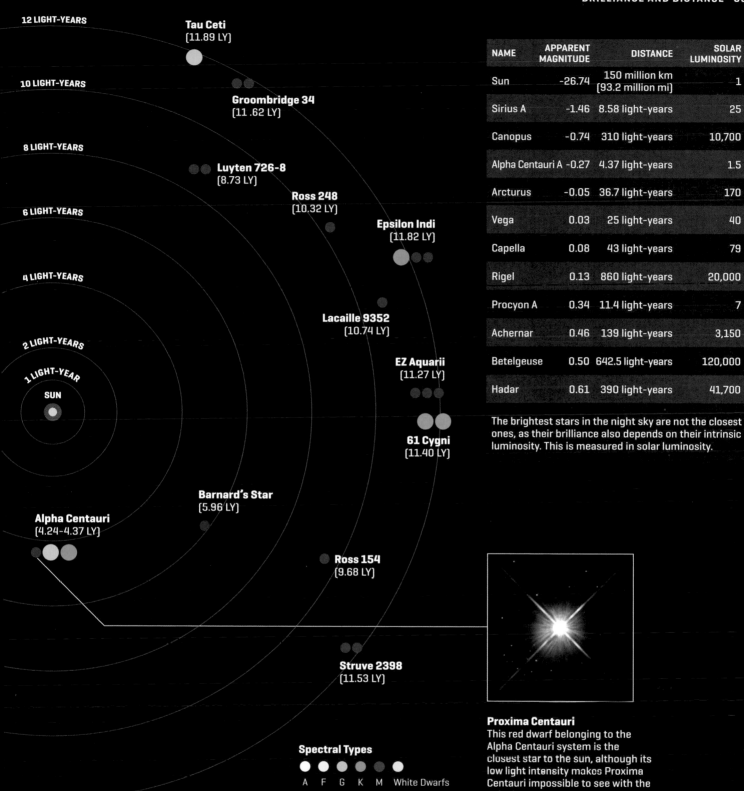

12 LIGHT-YEARS

Tau Ceti
[11.89 LY]

10 LIGHT-YEARS

Groombridge 34
[11 .62 LY]

8 LIGHT-YEARS

Luyten 726-8
[8.73 LY]

Ross 248
[10.32 LY]

6 LIGHT-YEARS

Epsilon Indi
[11.82 LY]

4 LIGHT-YEARS

Lacaille 9352
[10.74 LY]

2 LIGHT-YEARS

1 LIGHT-YEAR

SUN

EZ Aquarii
[11.27 LY]

61 Cygni
[11.40 LY]

Barnard's Star
[5.96 LY]

Alpha Centauri
[4.24-4.37 LY]

Ross 154
[9.68 LY]

Struve 2398
[11.53 LY]

NAME	APPARENT MAGNITUDE	DISTANCE	SOLAR LUMINOSITY
Sun	-26.74	150 million km [93.2 million mi]	1
Sirius A	-1.46	8.58 light-years	25
Canopus	-0.74	310 light-years	10,700
Alpha Centauri A	-0.27	4.37 light-years	1.5
Arcturus	-0.05	36.7 light-years	170
Vega	0.03	25 light-years	40
Capella	0.08	43 light-years	79
Rigel	0.13	860 light-years	20,000
Procyon A	0.34	11.4 light-years	7
Achernar	0.46	139 light-years	3,150
Betelgeuse	0.50	642.5 light-years	120,000
Hadar	0.61	390 light-years	41,700

The brightest stars in the night sky are not the closest ones, as their brilliance also depends on their intrinsic luminosity. This is measured in solar luminosity.

Proxima Centauri
This red dwarf belonging to the Alpha Centauri system is the closest star to the sun, although its low light intensity makes Proxima Centauri impossible to see with the naked eye.

Spectral Types

A   F   G   K   M   White Dwarfs

Inner Edge of the Oort cloud
11.5 light-days
300 billion km [186.4 billion mi]

Outer Edge of the Oort cloud
1 light-year 9.46 billion km [5.88 billion mi]

SIZE AND SCALE

The universe contains stars of all sizes, from extremely compact white dwarfs to supergiant stars whose diameters are 1,000 times that of the sun.

The interior of the sun could hold more than a million and a half planet Earths. Even so, our sun is a modest size star. There are some stars in the universe, such as supergiants and hypergiants, that have diameters hundreds of times larger than that of the sun. At the other extreme, smaller stars such as red dwarfs (Proxima Centauri), brown dwarfs, and white dwarfs (IK Pegasi B) have sizes closer to that of our solar system's planets.

Volume and Mass

A star's size varies over its lifetime. The sun, for example, will become a red giant at the end of its lifetime, growing so much that it will come into Earth's orbit. However, how long a star lives does not depend as much on its volume as on its mass, or the amount of material it contains. The red supergiant Betelgeuse, for example, is 1,000 times larger than the sun but only 20 times more massive. Spica is a relatively small star (roughly twice the size of the sun), but very massive, containing 10 solar masses.

Aldebaran
× 44

Rigel A
× 78

Wezen
× 200

Betelgeuse
× 1,000

UY Scuti
× 1,700

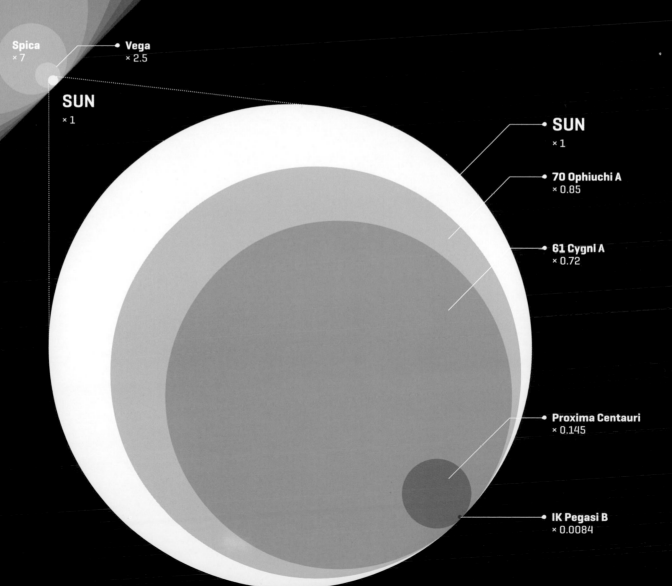

Arcturus
× 25

Spica
× 7

Vega
× 2.5

SUN
× 1

SUN
× 1

70 Ophiuchi A
× 0.85

61 Cygni A
× 0.72

Proxima Centauri
× 0.145

IK Pegasi B
× 0.0084

THE ENERGY OF STARS

A star's energy comes from nuclear fusion that occurs within it, creating temperatures ranging from tens to billions of degrees.

A star creates an enormous amount of energy, which starts in its core and passes through its outer layers, much of which is released into space as electromagnetic radiation. Fusion is the joining of two or more atomic nuclei to form a heavier nucleus. To kick off fusion in the star's interior, it must overcome the electromagnetic repulsion that exists between atomic nuclei, which requires high temperatures and crushing gravity.

Fusion Reactions

The most common fusion process in stars creates a helium-4 nucleus (containing two protons and two neutrons) from four hydrogen nuclei (four protons), as illustrated below. During this process, energy is generated in the form of gamma rays (photons) and subatomic particles (positrons and neutrinos), known as a proton–proton chain reaction, which is what occurs in most stars in the universe. More massive stars experience a carbon–nitrogen–oxygen (CNO) cycle: four protons generate a helium-4 nucleus, emitting two positrons, two neutrinos, and energy, a process in which the nuclei of carbon, oxygen, and nitrogen play a role in creating chain reactions.

From Hydrogen to Helium

This type of nuclear fusion, also called a proton–proton chain, is typical for stars found in the main sequence of the H-R diagram, such as the sun.

● Proton
  Neutron

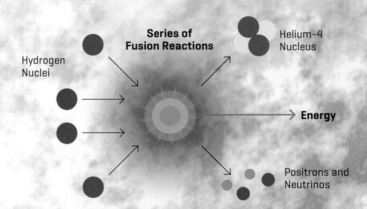

Hydrogen Nuclei

Series of Fusion Reactions

Helium-4 Nucleus

Energy

Positrons and Neutrinos

ENERGY'S TRIP

Much of the energy produced in a star's core begins to move outward, traveling through its layers over thousands of years to be released into the atmosphere.

CNO Cycle

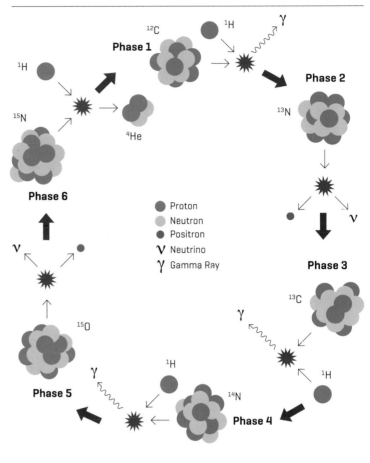

First, a carbon atom fuses with a hydrogen atom, forming nitrogen and releasing gamma rays. The resulting atom then loses particles and incorporates new hydrogen atoms to produce a helium atom. The leftover carbon allows the cycle to start again.

MASS VERSUS GRAVITY

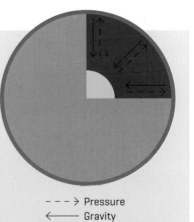

The life of a star requires a delicate and dynamic balancing act. On the one hand, its enormous mass forces the star to contract. On the other, its nuclear fusion reactions release energy that counteracts this contraction. During a star's lifetime, its equilibrium is thrown off as the star exhausts its nuclear fuels. As it increases in pressure and temperature to allow it to use new fuels and fusion reactions, the star will temporarily regain its balance.

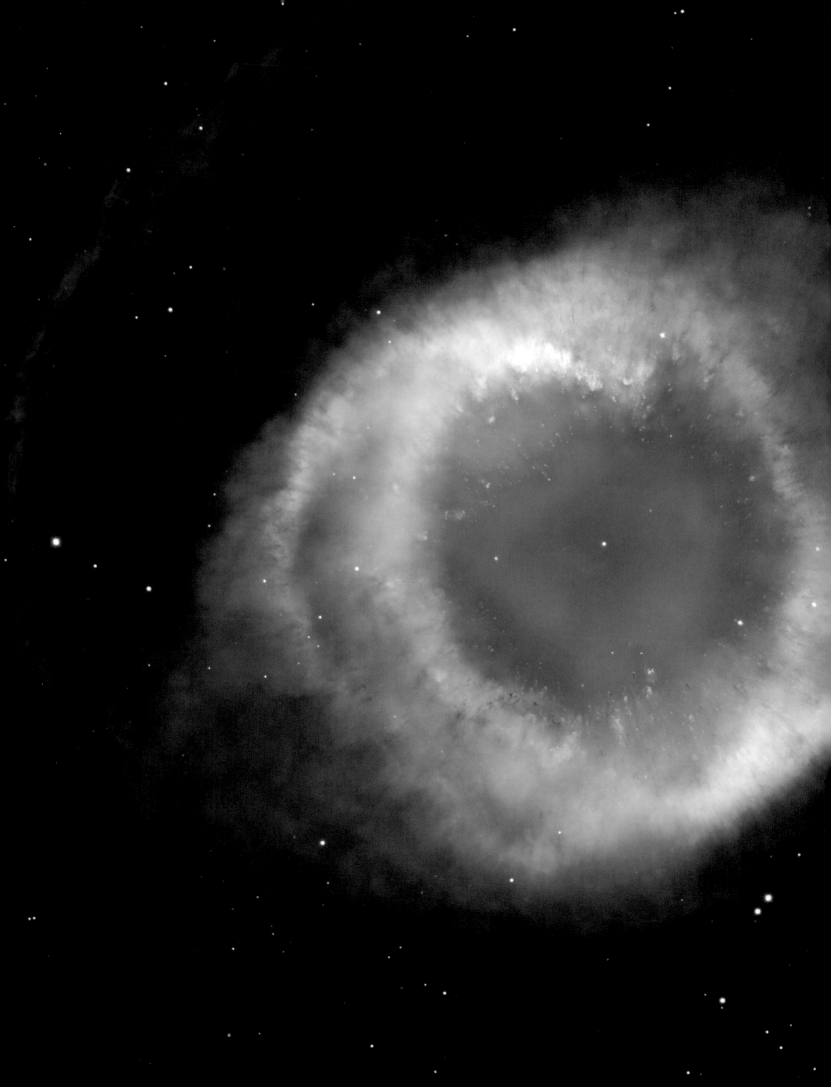

THE EVOLUTION
OF STARS

At the end of their lives, many stars enter the red giant phase, eventually ejecting their external layers to become a beautiful planetary nebula like the Helix, while its core contracts and becomes a white dwarf.

THE LIFE CYCLE OF STARS

A star's mass when it is born determines its fate. It will either burn through its fuel and be consumed, like dwarf and medium stars, or it will collapse violently. The more massive the star, the hotter and brighter it is, burning so hot that its life will be over much faster than that of smaller stars.

Stars Larger than the Sun
High-mass stars live only a few million years. Their frantic rates of nuclear fusion cause them to become enormous supergiants that end up collapsing in a catastrophic and extraordinarily violent explosion called a supernova, which can burn so bright that it outshines billions of stars in its galaxy. When it fades, all that is left is an extremely dense neutron star or a black hole.

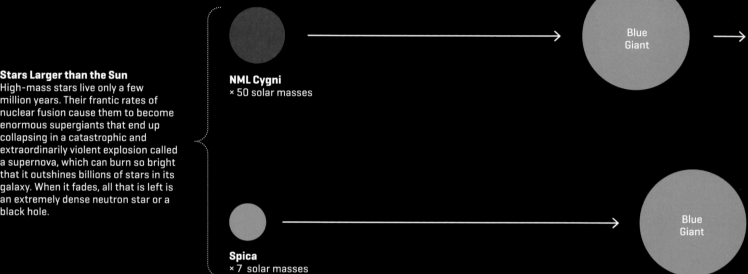

NML Cygni
× 50 solar masses

Blue Giant

Spica
× 7 solar masses

Blue Giant

Stars Like the Sun
Stars similar in mass to the sun stay in the main sequence of the H-R diagram for billions of years, fusing hydrogen into helium. They eventually become cold red giants, swelling to over 100 times their original size. The star slowly shuts down, its core contracting into a white dwarf while its external layers of gas are ejected as a planetary nebula.

Sun
× 1 solar mass

Stars Smaller than the Sun
Very low mass stars such as red dwarfs are the ones that live longest. Convection, which is dominant in stars that have less than 40 percent of the

Proxima Centauri
× 0.123 solar mass

How long a star lives depends primarily on its mass, which determines how quickly it burns through its fuel by way of nuclear fusion. Low-mass main sequence stars, like our sun, spend most of their lives in hydrostatic equilibrium: they burn hydrogen into helium steadily, keeping a balance with the inward force of gravity, until their fuel runs out. But high-mass stars have a higher luminosity, causing them to burn through their fuel much more quickly. Very low-mass stars last the longest: red dwarfs burn through their hydrogen stores very slowly, which means they can live for a long time.

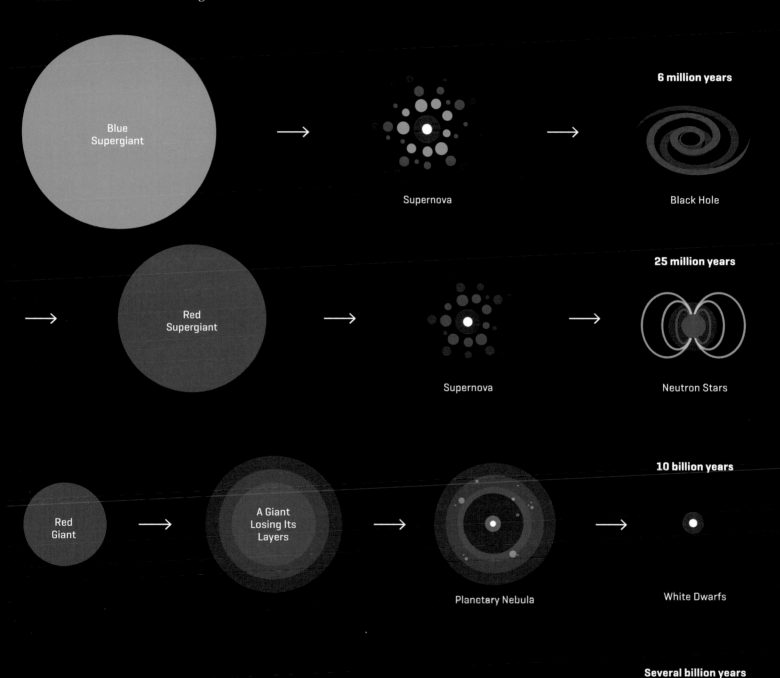

6 million years

Blue Supergiant → Supernova → Black Hole

25 million years

Red Supergiant → Supernova → Neutron Stars

10 billion years

Red Giant → A Giant Losing Its Layers → Planetary Nebula → White Dwarfs

Several billion years

Black Dwarf

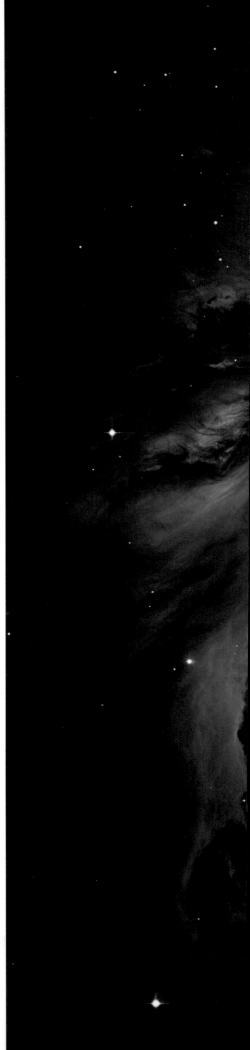

NEBULAE

The seemingly empty space between stars, known as the interstellar medium, is filled with gas and dust—the primary material needed to create new generations of stars.

Nebulae are immense clouds of gas and dust located in the interstellar medium, and they are especially abundant in the arms of spiral galaxies such as the Milky Way. The majority of nebulae are invisible and can only be detected because they block the passage of light from objects farther away. They can become visible when nearby stars illuminate them, creating peculiar and spectacular shapes.

Types of Nebulae

There are four types of nebulae. The first, called dark or absorption nebulae, are only visible because they do not allow light from other objects through, including other nebulae. The second type, reflection nebulae, are luminous because their dust particles reflect and scatter starlight the same way Earth's atmosphere scatters the sun's light. The third type, emission nebulae, emit their own light produced by young stars that formed in their interior. They are the largest and most striking, giving off light in a variety of colors. The fourth type, planetary nebulae, are the ghostly final remnants of a low-mass star, formed by ejected gas wrapped around the star's naked core, or white dwarf.

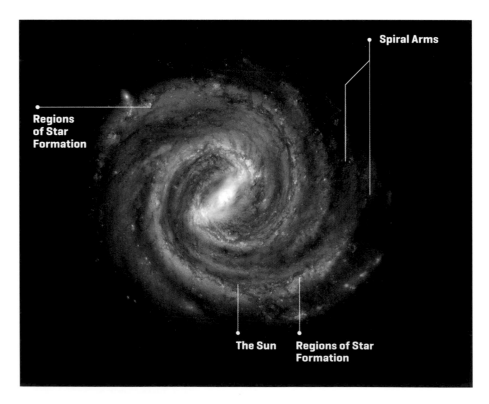

Regions of Star Formation

Spiral Arms

The Sun   **Regions of Star Formation**

REGIONS OF STAR FORMATION
The arms of spiral galaxies, such as those of the Milky Way, are believed to favor the creation of zones of high-density gas and dust, which are perfect environments for stars to form.

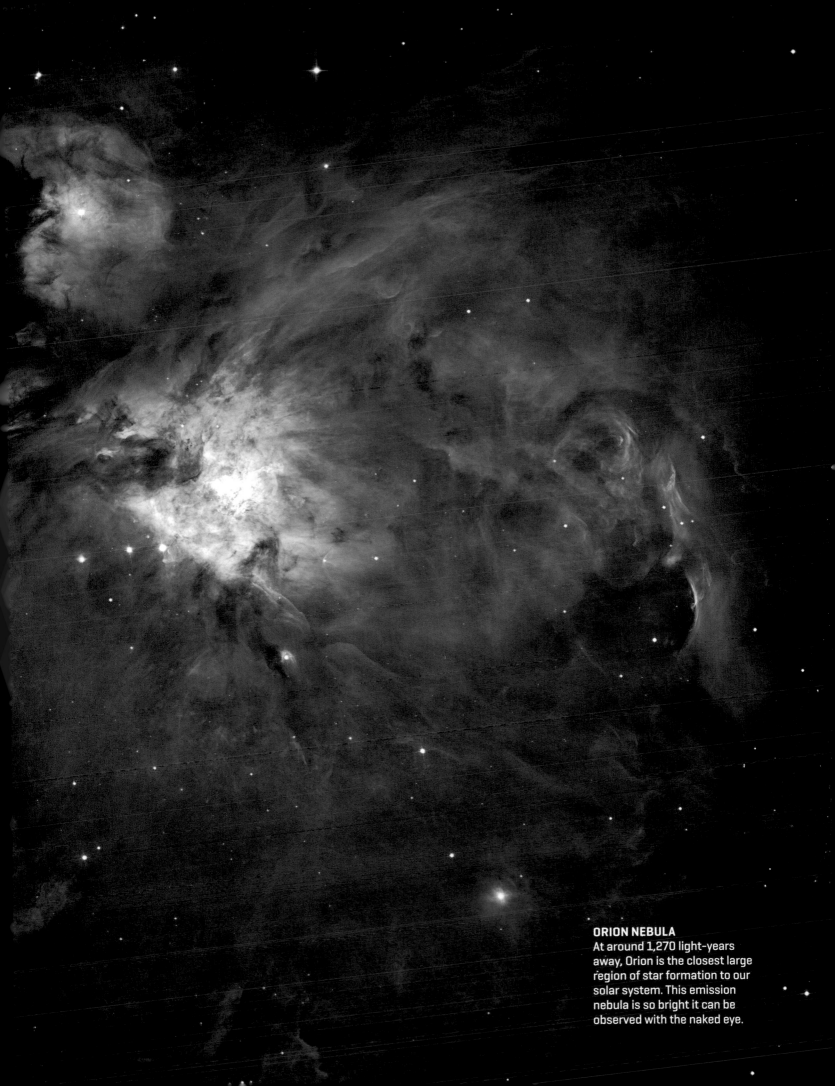

ORION NEBULA
At around 1,270 light-years away, Orion is the closest large region of star formation to our solar system. This emission nebula is so bright it can be observed with the naked eye.

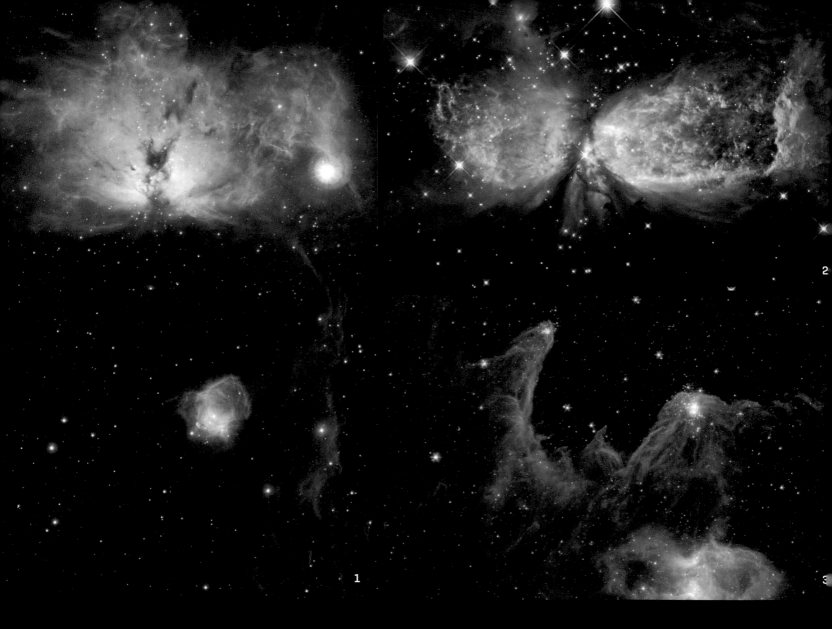

EMISSION NEBULAE

Flame Nebula

1 This region of star formation is located only 1,400 light-years away from Earth in Orion's Belt. Its characteristic, flame-shaped light could be due to Alnitak, an extremely hot and bright star: it is 100,000 times more luminous than our sun and some 20 times its mass.

Celestial Snow Angel

2 This high concentration of gas and dust has a large, massive star at its center. Molecules of interstellar material reflect its dim light, allowing us to observe the shock waves of its gases. Twin lobes of super-hot gas, glowing blue, stretch outward from its central star like the wings of a large, celestial angel.

Soul Nebula

3 These spectacular towers of gas are located 7,000 light-years away. Nicknamed "the mountains of creation," the nebula is shaped by the radiation of very hot and massive stars. It is believed that its powerful stellar winds (fast-moving flows of protons, electrons, and atoms of heavier metals ejected from the star) have compressed parts of the cloud and activated a second surge of star formation.

Carina Nebula

4 This image illuminates the filaments of gas that have escaped the large cloud, ejected due to the pressure of stellar winds from recently formed stars in its interior. With time, the young stars will stop the star-forming process, dissipating the dust and gas from the nebula.

Tarantula Nebula

5 This immense complex is located some 163,000 light-years from Earth in the Large Magellanic Cloud, a satellite dwarf galaxy of the Milky Way. It contains some 800,000 stars and protostars and is one of the largest known regions of star formation.

Lagoon Nebula

6 Located some 5,000 light-years from Earth toward the center of the Milky Way, the brightest part of this nebula is known as the Hourglass, which has a high rate of star formation. Its curious form is the result of extreme stellar winds and intense light from young stars.

STAR FORMATION

In the transformation from gas and dust to
newly formed star, gravity is the only factor
that really matters.

The creation of stars begins in the
heart of cold molecular clouds that are
abundant in the spiral arms of galaxies.
When stars die, their atmospheres
return to space, their elements enriching
the stellar gas and dust. Nebulae
begin to collapse, compressed by
shockwaves from a galaxy collision or
a nearby supernova, breaking up into
denser zones where these materials

begin to clump together. The clumps
grow, drawing in stellar material and
forming discs that may one day become
planetary systems. After an unstable
childhood characterized by increases
in gravitational force and temperature,
these star embryos gather enough heat
and mass to burn in the form of nuclear
fusion, fighting the gravitational forces
that are always pushing against them.

Moving Nebulae
The nebulae starts to
collapse under mutual
gravity, triggered by some
kind of gravitational
disturbance that breaks
it into dense clumps of
swirling gas and dust.

Protostars
Gravitational collapse
continues over tens of
thousands of years as the
protostar gains mass and
heats up. Shocked gas
from the inner edge of
the disc continues to fall
inward, and is emitted in
large jets at either end of
the protostar.

Gravitational Collapse
The temperature at the
center of these clumps
increases as a clump
starts to collapse under
its own gravitational
attraction, speeding up
to create a large, rotating
gas disc, and a protostar
begins to take shape.

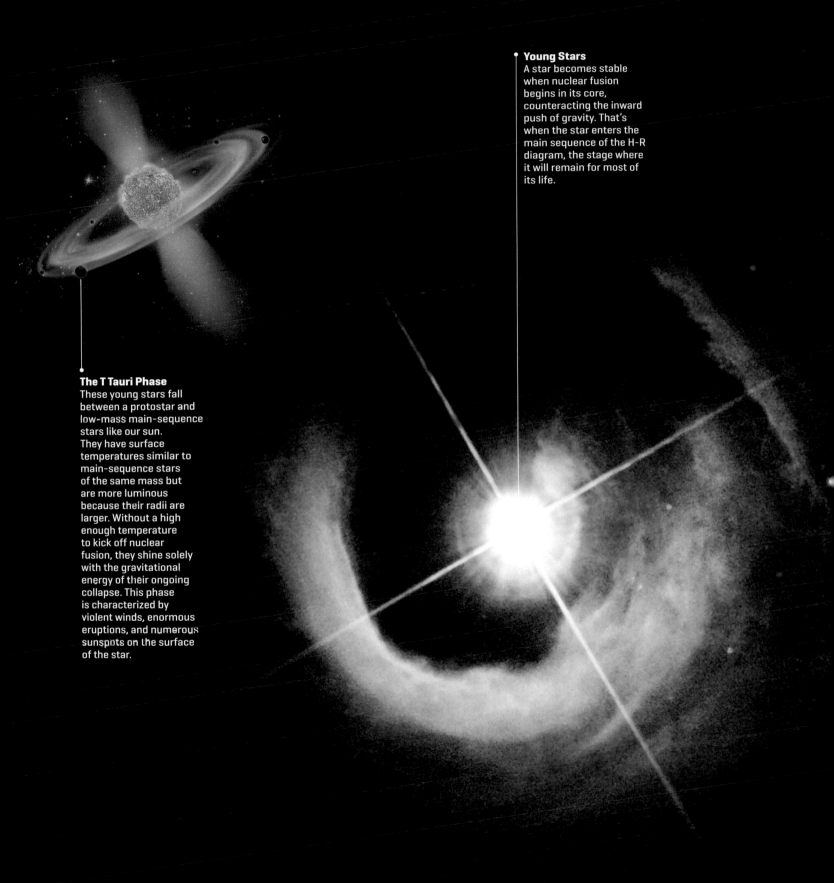

Young Stars
A star becomes stable when nuclear fusion begins in its core, counteracting the inward push of gravity. That's when the star enters the main sequence of the H-R diagram, the stage where it will remain for most of its life.

The T Tauri Phase
These young stars fall between a protostar and low-mass main-sequence stars like our sun. They have surface temperatures similar to main-sequence stars of the same mass but are more luminous because their radii are larger. Without a high enough temperature to kick off nuclear fusion, they shine solely with the gravitational energy of their ongoing collapse. This phase is characterized by violent winds, enormous eruptions, and numerous sunspots on the surface of the star.

YOUNG STARS

The start of fusion reactions in its nucleus marks the
passage to maturity for a star, a process that is a factor
in causing an increase in temperature.

A star's passage to maturity begins when fusion
reactions start to take place in its nucleus, one of
the factors that cause it to increase in temperature.
After the gas cloud that gave birth to the star breaks
up, its increased density and movement cause its
temperature to rise. At the same time, the star gains
speed, collecting material that forms a flat disc that
rotates around it. Nuclear fusion reactions are already
taking place in an element called deuterium, a heavier
form of hydrogen whose fusion produces relatively
low temperatures. But a rise in temperature triggers
the fusion of hydrogen to helium—the characteristic
fusion found in a star—that marks its official entrance
into the main sequence.

The Clouds Part
As the star shines, it continues to collect dust and gas,
which is ejected by stellar winds and is released in
jets by the star's new magnetic field. The rest of the
chemical compounds that surround it are vaporized by
the heat, and as the dust and gas begin to dissipate, the
star is revealed.

Angular Momentum
According to the law of
conservation of angular
momentum (the inertia
contained by rotating
objects), when a stellar cloud
gets smaller it increases its
speed, just as ice skaters
rotate faster when they pull
in their arms during a spin.

FROM COLLAPSE TO BREAKUP

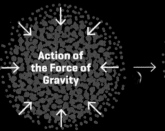

Action of
the Force of
Gravity

Phase 1
The gravitational attraction
between molecules and
grains of dust in the cloud
increase its density, which
in turn causes collapse.

Phase 2
Slight differences in density
inside the cloud make it
fragment as the cloud
collapses.

Phase 3
The fragments begin
to compress due to the
gravitational attraction
between their components,
creating a protostar.

Ejected Material
Young stars emit powerful jets of material perpendicular to the plane of rotation, helping to dissipate part of the rotational energy of the system and allowing the star to continue growing as it attracts material.

THE DISC OF RECENTLY FORMED STARS
In this illustration, the protostar has formed in the center of a rotating disc, and lumps around it have begun to grow by drawing in other material.

Gaining Mass
One of the clumps in the molecular cloud has collapsed. The material falls toward the center and becomes part of the star, which increases in mass. The motion energy transported by the material is converted into heat, increasing the temperature of the young star.

Traveling along the H-R diagram

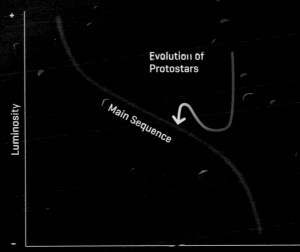

Evolution of Protostars

Main Sequence

Luminosity

Effective Temperature

When the interior of the young star reaches approximately 15 million degrees Celsius, the fusion of hydrogen to helium, characteristic of mature stars, begins. This represents admission to the main sequence in star formation.

Gas and Dust
In the protoplanetary disc, particles of dust gradually come together. The gas tends to be expelled toward the farthest parts of the disc due to stellar wind activity.

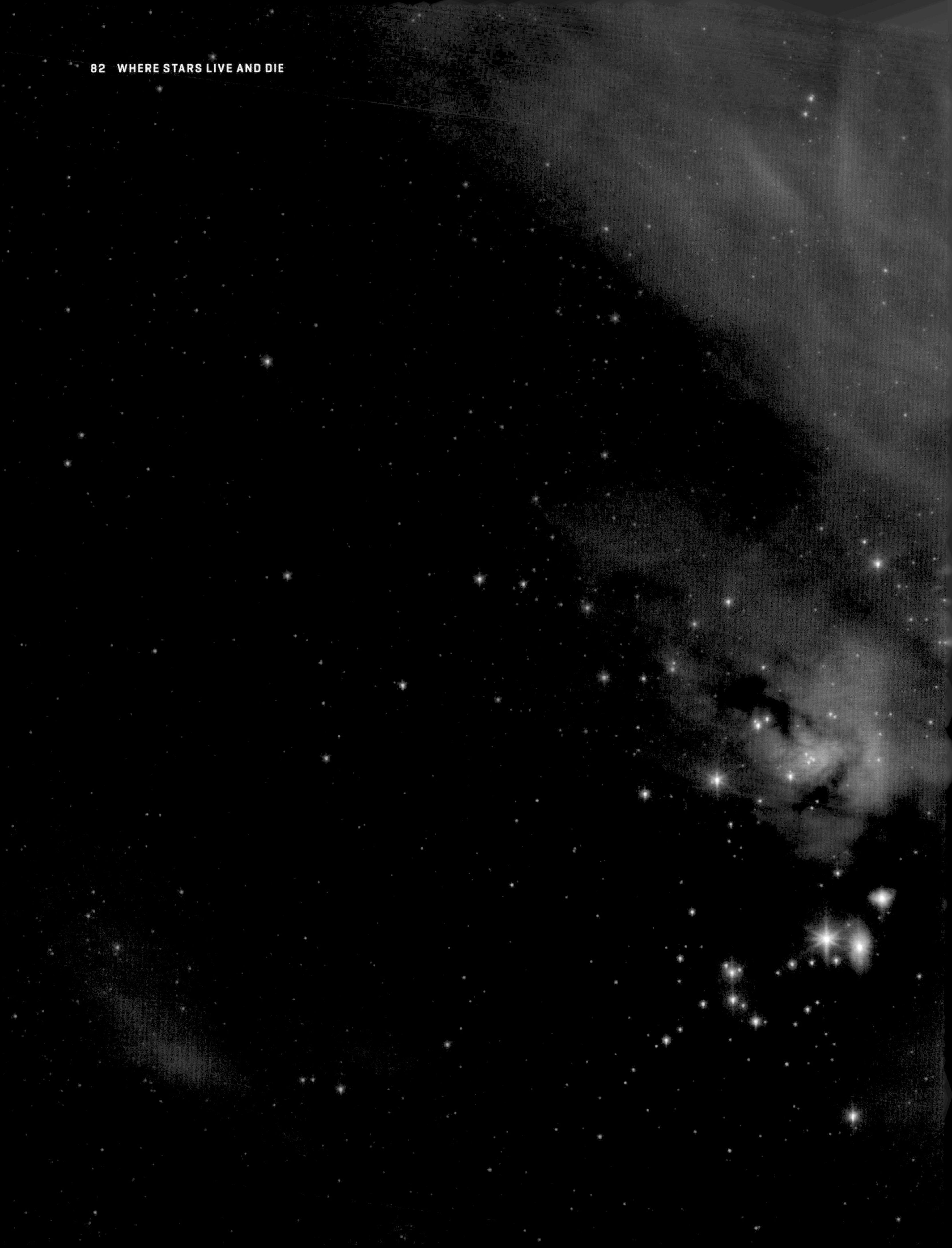

RHO OPHIUCHI

Young stars shine below the dust layer in Rho Ophiuchi, one of the regions of star formation closest to our solar system. In this image, young stars appear red because of the dust that surrounds them. They are hugged by gas discs that will go on to become future planetary systems. Other, more evolved stars appear blue, as the dust around them has already dissipated.

STAR CLUSTERS

Stars tend to be born in large groups, linked together by gravity until they scatter: these are called stellar clusters.

When molecular clouds of dust and gas collapse and break up into clumps, they give rise to a large number of stars. Eventually, the cloud that gave birth to them is lit up by the intense stellar winds that stars emit. These tight clusters of stars can number in the thousands. Open clusters have no special structure: they are linked by gravity until, slowly, the galaxy's momentum causes them to disperse. Globular clusters, by contrast, have a characteristic spherical shape and could be the remains of small galaxies.

Clustering on the Galactic Plane
In the Milky Way, open clusters are typically found on its disc, or galactic plane, the flat part where there is plenty of gas and dust for star formation in the spiral arms of the galaxy.

OPEN CLUSTERS

Open clusters contain anywhere from a few dozen stars to thousands of them, but their low density makes them vulnerable to internal collisions and external disruptions like a passing molecular cloud whose gravitational force could tear the cluster apart. That's why there are relatively few old open clusters; most are relatively young at only hundreds of millions of years old.

Cluster M25
This bright open star cluster contains millions of stars and is visible from Earth with the naked eye. Its blue light indicates young, hot stars.

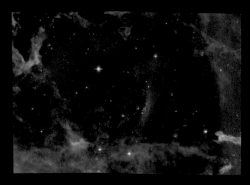

Cluster NGC 2244
Located in the Rosette Nebula, this cluster contains many O-type stars that emit a large amount of radiation.

CLUSTERS OF THE MILKY WAY

In galaxies like ours, there are star clusters that seem almost like independent galaxies. The densest of these globular clusters contain stars that are at least 10 billion years old, making them some of the oldest objects in the Milky Way.

Galaxy Satellites
In the Milky Way, globular clusters orbit around the galaxy. This, along with their shape and age, makes us suspect they could be the cores of minor galaxies that were absorbed by our own.

GLOBULAR CLUSTERS

These extremely dense groups of stars are made up of hundreds of thousands— sometimes even millions—of very old stars. Their low metal content suggests these stars are older than our sun, formed when the interstellar medium was poor in elements other than hydrogen and helium.

Cluster M15

M15 is one of 170 known globular clusters in our galaxy. It comprises more than 100,000 stars and is estimated to be some 12 billion years old.

Omega Centauri

This cluster orbiting our galaxy is the brightest in the sky. The stars at its center are bound very closely together: they are only 0.1 light-years apart.

RED GIANTS

After billions of years fusing hydrogen, stars eventually use up their fuel. They cool down and increase in size, turning into red giants.

When the cores of stars like our sun (low-mass stars between .08 and 8 solar masses) run out of hydrogen, helium becomes the main element. Because the star's temperature isn't high enough for a fusion reaction, the star can no longer fend off the crushing force of gravity. The star contracts, collapsing inward, bringing all additional hydrogen into a hot enough zone to kick off fusion again in the shell around the core.

The Final Fusion Reaction

The star's helium nucleus, under strong pressure from the fusion reactions that have burned through its surrounding layers, heats up. Upon reaching 100 million degrees, the star starts its last nuclear fusion reaction, converting helium to carbon through a process known as triple alpha.

Becoming a Giant

When the star uses the helium in its core, the interior starts to contract. The increase in temperature starts helium fusion in adjacent layers. The star now has a carbon-oxygen core, surrounded by layers where the fusion of hydrogen and helium continues for a few more million years. The star swells and expands as its surface cools, becoming a red giant.

THE TRIPLE ALPHA PROCESS

Red giants produce the vast majority of carbon in the universe through a fusion process called the triple alpha process. When hydrogen fusion ceases, the star finds a new way to produce energy: the almost simultaneous collision of three helium-4 nuclei (^{4}He, also called alpha particles), which merge to give rise to a carbon-12 (^{12}C) nucleus through intermediate beryllium-8 (^{8}Be) formation and generate energy in the form of gamma rays.

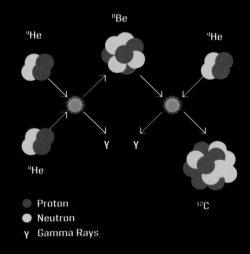

^{8}Be

^{4}He ^{4}He

γ   γ

^{4}He

^{12}C

● Proton
● Neutron
γ Gamma Rays

R SCULPTORIS

The ALMA radio telescope,
located in the Chilean Atacama
Desert, has captured a
spectacular view of the star
R Sculptoris, a red giant
surrounded by its expelled
gas. The curious spiral shape
of the gas could be due to the
presence of a companion star.

Traveling Along the H-R diagram

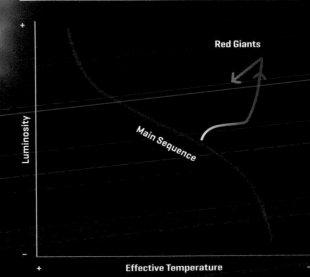

Red Giants

Luminosity

Main Sequence

Effective Temperature

A star leaves the main sequence when it runs out of
fuel, going along the path of red giants. Its surface
temperature lowers while its luminosity increases, both
a product of its large size. Once helium fusion has begun,
the star contracts again, and its surface warms back up.

THE SUN AS A RED GIANT
When our sun eventually becomes a
red giant, billions of years from now, its
expansion will be extraordinary: it will
become 150 times larger than its current
size and devour Earth. Our home planet
will already have burned by the time it
reaches us, becoming a wasteland where
no water exists and oceanic beds like this
one can be seen clearly

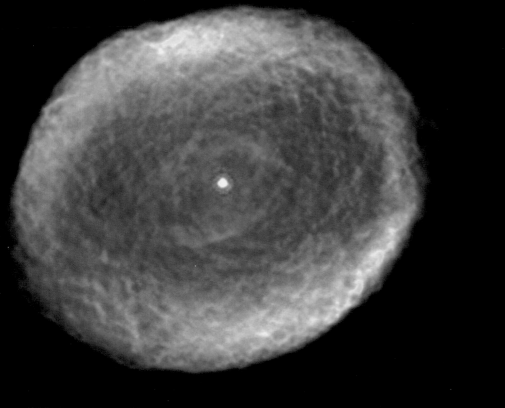

SPIROGRAPH NEBULA
The IC418, or Spirograph Nebula, has the remains of a red giant star in its center, surrounded by a striking shroud of geometric shapes.

RED SPIDER NEBULA
The Red Spider Nebula is a spectacular example of what happens when a white dwarf's extreme heat and powerful stellar winds carve whimsical shapes in clouds of gas.

PLANETARY NEBULAE

When a red giant's life comes to an end, it expels its exterior layers, leaving behind the naked core surrounded by an expanding envelope of ionized gas in the form of a planetary nebula.

Eventually, the nuclear reactions inside a red giant come to an end. In an attempt to stay alive, the star will start uncontrolled helium fusion in the layers that, until this point, have been fusing hydrogen. The layers surrounding the giant's core are gradually expelled, forming a ghostly planetary nebula. The star's carbon core remains in the center, becoming a very dense, hot star, called a white dwarf. Its ultraviolet radiation ionizes the nebula around it, causing it to shine spectacularly.

Planet-like Spheres
First discovered in the 18th century, the spherical, symmetrical shape of these nebulae led astronomers to call them "planetary" due to their resemblance to the planets as seen through the telescopes of that era. However, the only thing the two have in common is their name.

CAT'S EYE NEBULA
The layers surrounding the core of a dying star are expelled at high speeds, creating beautiful nebulae like the Cat's Eye, seen here as captured by Hubble.

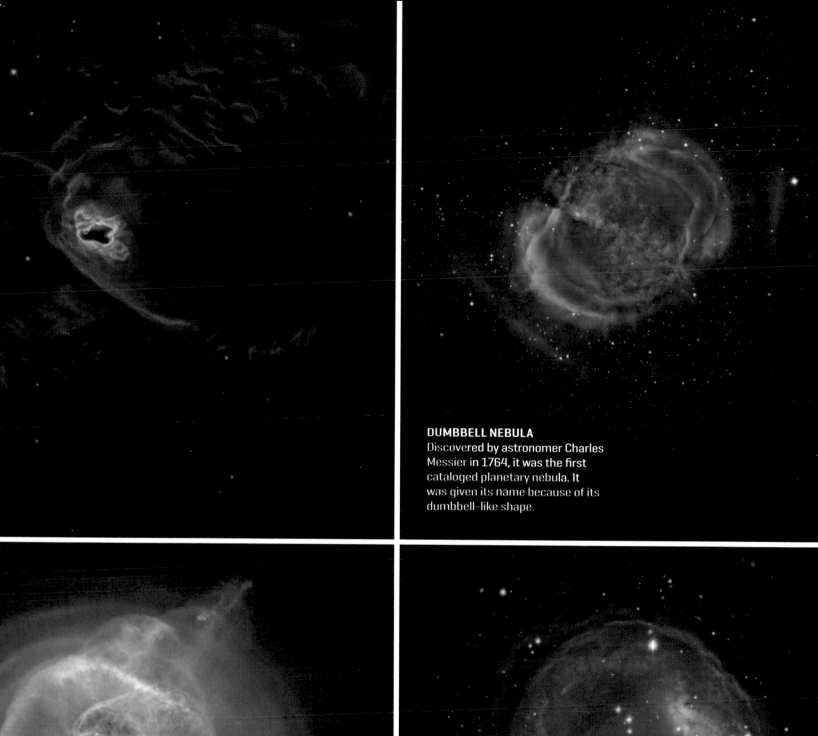

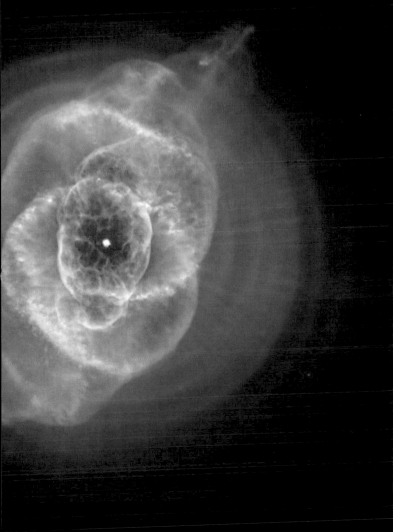

DUMBBELL NEBULA
Discovered by astronomer Charles Messier in 1764, it was the first cataloged planetary nebula. It was given its name because of its dumbbell-like shape.

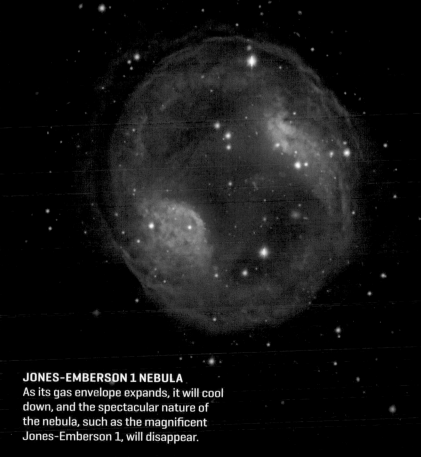

JONES-EMBERSON 1 NEBULA
As its gas envelope expands, it will cool down, and the spectacular nature of the nebula, such as the magnificent Jones-Emberson 1, will disappear.

WHITE DWARFS

Most stars, once they have shed their external layers and are left to the mercy of gravity, cool down and languish over billions of years, ultimately turning into white dwarfs.

A white dwarf is the heart of what used to be a star, with an initially high temperature on the order of 100,000°C (180,000°F) and an enormous density. In these conditions, the material within the star is in a state called "degenerate," where the atoms' electrons resist compression, as seen in the figure below. It still has the mass of the sun, but it is no bigger than Earth and is so dense that just a teaspoon of it would weigh five tons. The fate of a white dwarf is to cool off completely and become a cold,

invisible body. However, the loss of heat is so extraordinarily slow that scientists posit the universe hasn't existed long enough for any white dwarf to cool completely.

The Oldest Diamonds in the Cosmos

As a white dwarf slowly cools, its material can crystallize. The oldest white dwarfs in the universe, which have the lowest temperatures, are made of carbon crystalline structures not unlike an enormous diamond.

The Pressure of Degenerated Material

In degenerated gas, pressure has compressed the electrons into their least energetic state. The laws of quantum mechanics prevent more than two electrons from occupying the same energy level, and gravity cannot compress the dwarf any further. Thus white dwarfs survive not by internal fusion but by quantum mechanical principles, which save it from complete collapse.

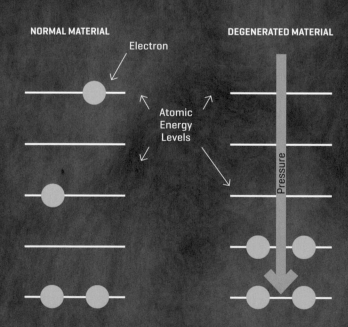

NORMAL MATERIAL

Electron

Atomic Energy Levels

DEGENERATED MATERIAL

Pressure

SMALL STARS WITH A LARGE MASS
This illustration shows a white dwarf
passing in front of a red dwarf, its
gravity distorting the light from the
larger star—an example of just how
large a mass these small stars have.

Comparison Between a White Dwarf and Earth

	DIAMETER	MASS	DENSITY	GRAVITY
White Dwarfs	10,000 km	2×10^{30} kg	2×10^{6} g/cm³	10⁶ m/s²
Earth	10,000 km	6×10^{24} kg	6 g/cm³	10 m/s²

MASSIVE STARS AND STAR SYSTEMS

The Crab Nebula is the remains of a supernova observed in 1054 by Chinese and Arab astronomers. According to estimations, the original star would have equaled between eight and 12 solar masses.

THE SHORT LIFE OF A MASSIVE STAR

The most massive stars shine intensely but use up their fuel relatively quickly. That's why their existence is limited to a few million years.

To keep hydrostatic equilibrium, where the pressures inside a star balance out the force of gravity, a massive star has to create nuclear fusion at a frenzied rate. But gravity is relentless, and new cycles of reactions occur as the temperature in the interior increases.

The Progress of Fusion

When a certain fuel runs out in the heart of a star, the nuclear fusion reaction that used that element is extinguished, and what remains is relegated to an adjacent layer. The core contracts under the immense weight of the star, and the increase in temperature allows it to start a new fusion reaction. In this way, the star's structure stratifies in layers where different nuclear fusion reactions happen simultaneously (as can be seen in the diagram below). They stay that way as long as the layers actively continue the process of fusion. After hydrogen and helium fusion, carbon fusion occurs, giving rise to oxygen, neon, sodium, magnesium, and silicon. These elements also undergo fusion in a cascade of countless reactions that are less energetic and shorter each time, which is more inefficient for the star. With each new cycle of reactions, the star tries to gain time in its desperate fight to maintain hydrostatic equilibrium.

The Layers of Massive Stars
The interior of old, massive stars has a layered structure characterized by the dominance of different chemical elements.

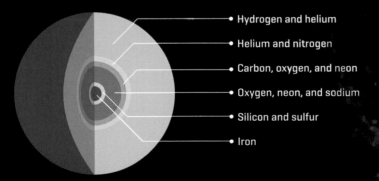

- Hydrogen and helium
- Helium and nitrogen
- Carbon, oxygen, and neon
- Oxygen, neon, and sodium
- Silicon and sulfur
- Iron

NUCLEAR FUSION PROCESS	CENTRAL TEMPERATURE (K)	CENTRAL DENSITY (kg/m³)	DURATION
Hydrogen fusion (H ▸ He)	37 million	3,800	7.3 million years
Helium fusion (He ▸ C + O)	180 million	620,000	660,000 years
Carbon fusion (C ▸ Ne)	720 million	640 million	165 years
Neon fusion (Ne ▸ Mg + Si)	1.4 billion	3.7 billion	1.2 years
Oxygen fusion (O ▸ Si)	1.8 billion	13 billion	6 months
Silicon fusion (Si ▸ Fe)	3.4 billion	110 billion	1.5 days

Phases of nuclear fusion of a star with 25 solar masses. This shows the main elements that occur in reactions and other relevant circumstances during each phase.

ETA CARINAE

This star in the Carina constellation is located 7,500 light-years from Earth and is one of the largest known stars, between 100 and 150 times as massive as the sun. Surrounding it are the remains of the ancient eruptions that formed the Homunculus Nebula.

Traveling Along the H-R Diagram

During its short life, a massive star moves on the H-R diagram with each new fusion cycle. The swings correspond with the expansion of its external layers, the progressive loss of mass due to its powerful stellar winds, and the temperature changes that occur in its core with the start of new nuclear fusion reactions.

NEAR THE END OF THE STAR'S LIFE

Some of the massive stars found in the universe, such as Eta Carinae and WR 124 (above), are very close to becoming supernovae. The clouds that surround them are their outer layers, removed by their powerful stellar winds.

SUPERNOVAE

At the end of their lives, the most massive stars give way to gravity and explode in so-called supernovae.

A star stops creating elements when it gets to iron. Until it reaches that point, all of its fused elements are an energy source that allows it to maintain its hydrostatic equilibrium. Iron fusion, however, does not give off energy, beginning a battle that the star cannot win. Without a source of energy to maintain its equilibrium, the core becomes unstable and can't counteract its own gravity. Not all stars are left to this fate, only those whose core exceeds 1.4 solar masses (a number known as the Chandrasekhar limit). This process takes only a fraction of a second: electrons collide with protons, creating neutrons, neutrinos, and gamma rays in a catastrophic collapse. This phenomenon creates a giant shock wave moving at supersonic speeds through the star's outer layers, scattering the remains of the star. Though the star dies, new ones will be created out of the elements it fused during its lifetime.

The Destruction of a Giant Star

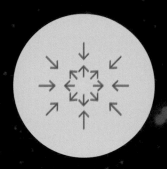 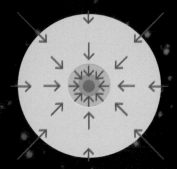 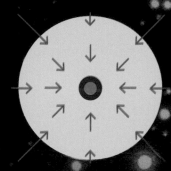 

1. The Resistance of Stars
The star's core withstands gravity's pressure, as it contains electrons that resist being compressed.

2. The Star Cannot Withstand Its Weight
When its core grows above 1.4 solar masses, the star's structure collapses.

3. The Collapse of a Star
The star falls in on itself at approximately the speed of light.

4. The Explosion
Shock waves from what remains of the star are launched outward in the form of a massive explosion.

 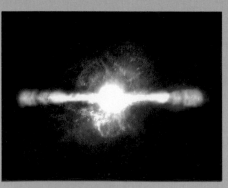

HYPERNOVAE

A superluminous supernova, or hypernova, occurs when an extremely massive star (more than 100 solar masses), like the Pistol Star (left), collapses. The death of these types of stars (right) has a luminosity that is at least 10 times more powerful than a normal supernova, and they create a rotating black hole when they collapse.

①

REMNANTS OF SUPERNOVAE

A remnant is a nebulous structure caused by the explosion of a supernova. It is made up of material released during the explosion along with interstellar material dragged along by the blast wave.

The explosion of a supernova releases a large amount of stellar material, creating a shock wave in front of the material it ejected, which, as it expands, heats the interstellar material. This material eventually heats up so much that it releases x-rays. Remnants are the main source of heavy chemical elements (especially oxygen). If the explosion of a massive star is still within the molecular cloud in which it formed, the remnant can cause interstellar gas near it to be compressed and to trigger the stellar formation process, resulting in more stars.

Stellar Material and Something More

As the outer layers of a star are shrugged off, leaving gases, which expand—as in the case of supernovae caused by gravitational collapse, which occurs in massive stars—the core contracts until it implodes and becomes a black hole or a neutron star, depending on its mass. On the other hand, thermonuclear supernovae, also known as Ia-type supernovae, disappear completely. This occurs in binary systems made of up a white dwarf and another star. The first captures the mass of its companion, and when it reaches critical mass, it explodes.

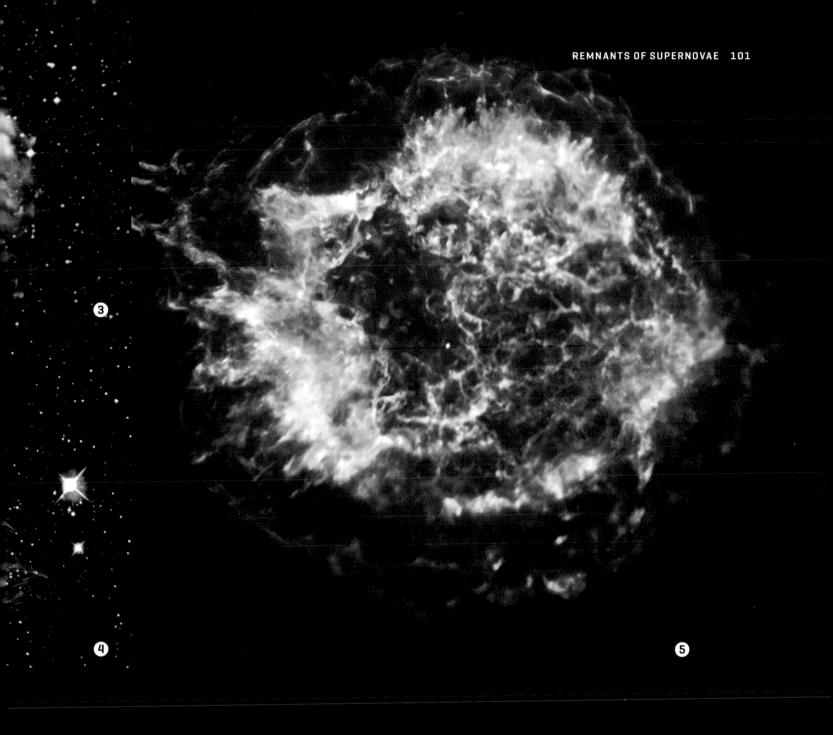

The Veil

1 The structure that results from a supernova is what is known as a supernova remnant. The Veil is the remnant of a massive star that exploded approximately 8,000 years ago. These photos obtained by the Hubble Space Telescope allow us to see a small part of it.

W49B

2 This colored image combines data from three observatories over three different wavelengths (x-rays, radio waves, and infrared) and is one of the best photographs ever taken of supernova remnant W49B, which could contain a black hole younger than the Milky Way. It is located 26,000 light-years from Earth and is thought to be approximately 1,000 years old.

RCW 103

3 At the end of the 20th century, a mysterious object was discovered in the heart of supernova remnant RCW 103, the gaseous remains of a very massive star that exploded more than two millennia ago. Found 10,000 light-years away from our solar system, the object is probably a pulsar: a neutron star that periodically releases radiation.

N49

4 This is the Greater Magellanic Cloud's brightest supernova remnant. Its gas filaments, which are the remains of a star that exploded around 5,000 years ago, are scattered across 30 light-years. N49, found some 163,000 light-years from Earth, probably contains neutron stars, according to measurements and data taken by the Chandra X-Ray Observatory.

Cassiopeia A

5 This supernova remnant is the strongest source of dazzling radio waves outside the solar system, found approximately 11,000 light-years away. Although no sources have confirmed the nature of the star before it became a supernova, it is estimated that its light first reached Earth three centuries ago.

NEUTRON STARS

After a massive star explodes as a supernova, its core can take one of two different evolutionary paths. If the star's nucleus has a mass less than three times that of the sun, gravitational contraction causes it to collapse and become a neutron star—an object with surprising properties.

A massive star's supernova explosion is one of the most spectacular phenomena in the cosmos. However, what happens in the depths of the supernova is even more extraordinary: as stellar material is ejected by giant shock waves, some of the universe's most enigmatic and intriguing objects begin to take shape in its interior—a black hole or a neutron star. Atoms in the star's core are compressed to become neutrons, which arrange themselves into a structure that can withstand the force of gravity: a dense ball of neutrons called a neutron star.

An Extraordinary Density
Made up mostly of neutrons—and with a mass higher than the sun's, compressed into a diameter of fewer than 10 kilometers (6.21 mi)—a neutron star is incredibly dense. One teaspoon of a neutron star's matter would weigh around a billion tons. With forces on its surface equivalent to a billion times that of our planet, a passing celestial body would need to be moving at least half the speed of light to escape the gravitational influence of such a massive body.

Pulsars: The Lighthouses of the Universe

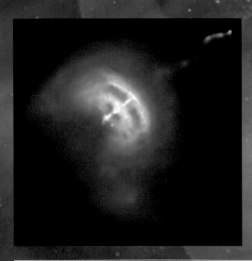

When a supernova has enough momentum and energy to blow outward, it sets its remaining core spinning in the opposite direction. As the core, now a neutron star, contracts, its rotational speed picks up, and when it spews jets of radiation along its magnetic field, its rotation makes them fan out like the beam of a lighthouse. This is a pulsar. When its jets are oriented toward Earth, the "lighthouse" effect can be seen blinking in sync with its rotation. This image shows the pulsar in the Veil, which was linked with a supernova remnant in 1968—evidence that supernovae can make neutron stars.

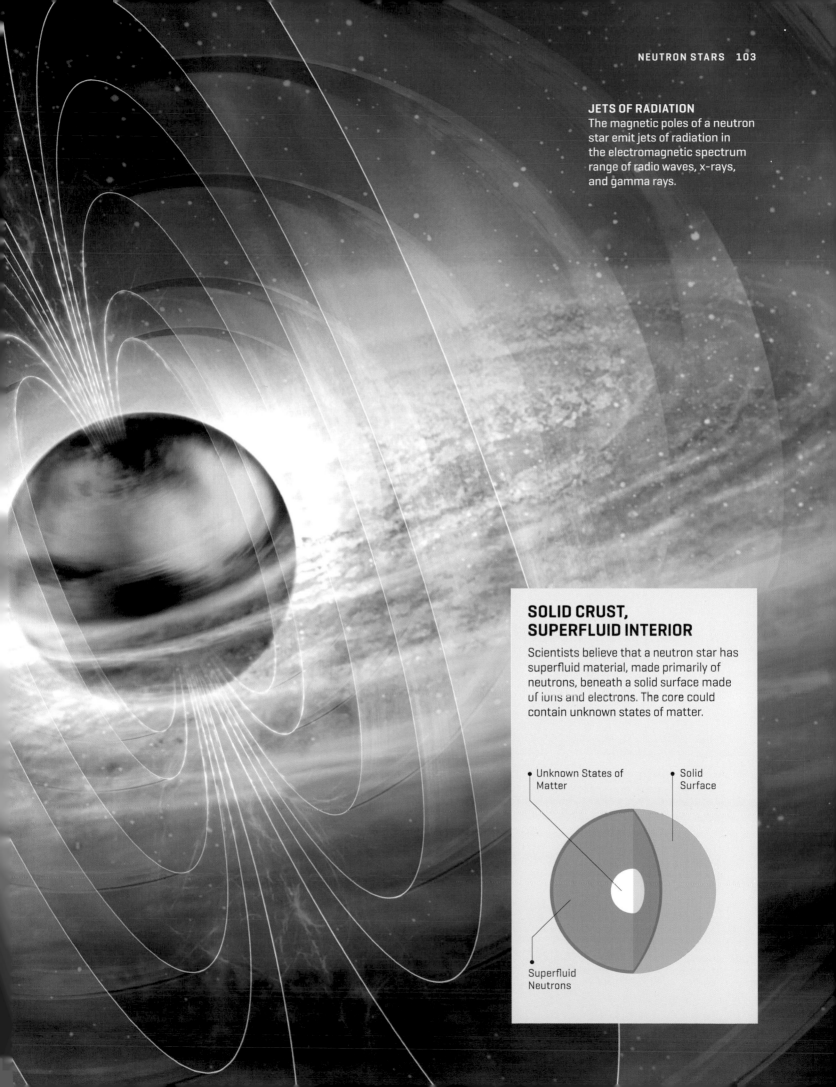

JETS OF RADIATION

The magnetic poles of a neutron star emit jets of radiation in the electromagnetic spectrum range of radio waves, x-rays, and gamma rays.

SOLID CRUST, SUPERFLUID INTERIOR

Scientists believe that a neutron star has superfluid material, made primarily of neutrons, beneath a solid surface made of ions and electrons. The core could contain unknown states of matter.

Unknown States of Matter

Solid Surface

Superfluid Neutrons

BLACK HOLES

These enigmatic objects are made of concentrated mass with a gravitational field so strong that it will consume anything that ventures too close.

If a post-supernova star's core is very large (more than three times the mass of the sun), it continues to shrink until gravity crushes it into an infinitely dense point in space. Matter as we know it ceases to exist. The result is a black hole, where gravity is so strong that nothing can escape its pull.

The Clue to Something Invisible

The event horizon is the boundary past which nothing can escape a black hole—not even light. As matter falls in, it heats up so much that it emits radiation before disappearing past the event horizon, helping scientists detect what is otherwise an invisible object. Although they are not yet sure what happens inside a black hole, scientists theorize that its center holds a gravitational singularity—a one-dimensional point that contains a huge mass in an infinitely small space, where space and time as we know them cease to exist.

The Structure of a Black Hole

A black hole is a region of space-time where, after crossing an event horizon, not even light can escape the force of its gravitational pull.

According to predictions based on current laws of physics, at the heart of a black hole lies a singularity: a one-dimensional point of infinite density that defies our notions of time and space.

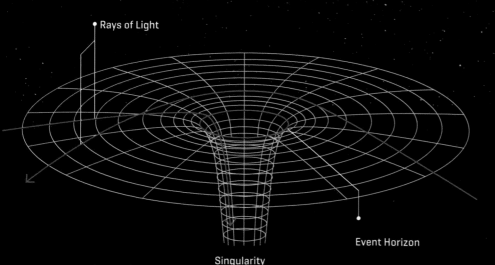

Rays of Light

Event Horizon

Singularity

SUPERMASSIVE BLACK HOLES

On April 10, 2019, the world got its first photo of a black hole (left). At the center of Messier 87, an elliptical galaxy in the Virgo cluster, an asymmetric ring of light visibly bends around a supermassive black hole. The photo was captured by the Event Horizon Telescope.

ROTATIONAL MOTION

This artist's depiction of a black hole includes its accretion disc, a structure made up of gas and dust that contributes to the mass of the central object.

BINARY STARS

The majority of stars come with companions, linked to each other by gravity and orbiting a common center of mass.

Large numbers of stars form in nebulae, and some of them get linked up by gravity and form what are called multiple star systems. The simplest is a binary system, made up of just two stars. The stars that make up multiple star systems can have very different masses, and therefore very different characteristics.

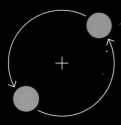

Identical Masses
If the mass of two binary stars is equal, the center of gravity is directly between the two.

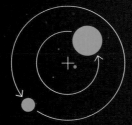

Different Masses
If the mass of two binary stars is different, the star with the higher mass will be closer to their shared center of gravity.

Large Differences in Mass
If the masses of two binary stars are very different, the center of gravity might be located within the more massive star.

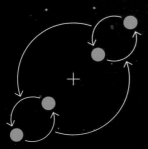

Quadruple System
If a system is made up of two binary star pairs, the center of gravity will be between the two systems.

KEPLER-35
This binary system is made of two stars very similar to the sun with relatively similar masses. Only 25 million kilometers (15.5 million miles) separate them, and they orbit each other every 21 days.

Eclipsing Binaries

Eclipsing binaries are binary systems whose orbital planes are oriented toward Earth, which allows for total or partial eclipses that we can see from Earth. These systems are extremely important in our quest to understand the stars—an analysis of their light curves and radial velocities lets scientists calculate key parameters, such as their mass and dimensions.

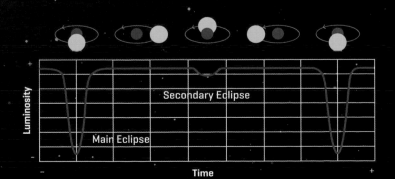

Luminosity

Secondary Eclipse

Main Eclipse

Time

STELLAR COEVOLUTION

The stars in star systems evolve, but it happens in several different ways. One of the stars in a pair might mature faster than the other, significantly altering their relationship to each other. In this illustration, such an imbalance in power is causing one star to take matter from its partner.

ALGOL

The Algol system has two stars with very different masses: Algol A [far right] is a B-type star that is much more massive than its companion Algol B [near right], a K-type star. This imbalance, and the fact that they are very close to each other at a distance of only 9 million kilometers [5.59 million miles], means that material continues to transfer between them.

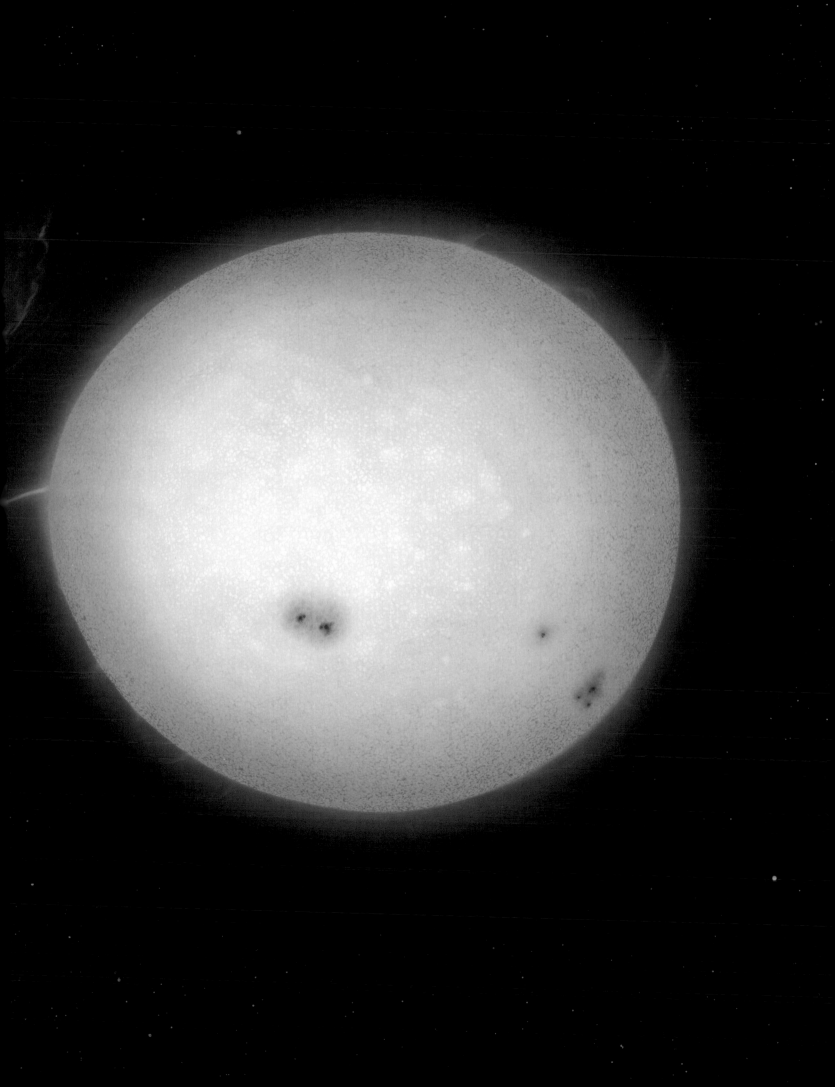

NOVAE

In binary systems in which one of the stars has become a white dwarf, acquiring a companion star's material can lead to violent energy emissions called novae.

As stars in a binary system evolve, one of them might become a white dwarf. The "normal" star in the pair will transfer material to the white dwarf, which begins to accumulate material on its surface. This outer layer of collected matter increases, amping up in pressure and temperature until violent, uncontrolled hydrogen fusion begins.

Stellar Explosion

This hydrogen fusion causes a burst of energy that temporarily increases the total brilliance of the system by hundreds or even thousands of times. The white dwarf is unaffected by the explosion, which affects only the external gas layers. This process can sometimes happen again in a star system, causing a new burst every so often.

The Explosion Process

The V959 Mon nova started with the ejection of material along the equator of the binary system. It collided with strong winds released from the poles, ejecting huge amounts of energy in the form of gamma rays (the red regions). In the last phase, the material was scattered into space, and the system returned to its calm state.

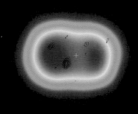
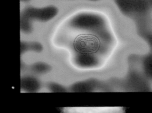
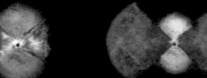

GK PERSEI

GK Persei was discovered at the beginning of the 20th century, when an increase in luminosity made it one of the brightest objects in the night sky. It has nova bursts every few years.

THE FIRST COMPLETE NOVA

On February 19, 1992, the European Space Agency captured a magnificent image of a thermonuclear explosion in a binary system in the constellation Cygnus, more than 10,000 light-years from Earth. It became the first nova ever observed from beginning to end.

OUR CURRENT VIEW OF THE STARS

The scientific and technological advances of the last few decades have revolutionized the way we observe the stars.

The vast size of the universe makes exploration beyond the solar system impossible. However, we can observe faraway objects by tracking their electromagnetic radiation, thanks to Earth-based and space telescopes. Since the 1990s, the pinnacle of space telescopes has been Hubble, whose images have been essential in validating theories about star formation, planetary birth, the galaxies' dynamics, and the makeup and expansion of the universe. Likewise, technologies like spectroscopy and asteroseismology have brought whole new aspects of the stars to light.

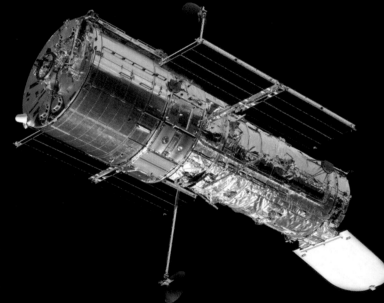

Asteroseismology

We may not be able to reach the interior of any stars, but analyzing the oscillations within them provides us with information about what they contain. This image illustrates how asteroseismology investigates the insides of stars by studying their pulsation modes, with a simulation of acoustic wave motion shown in blue and red.

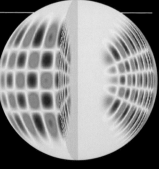

THE HUBBLE TELESCOPE
Since put into orbit in 1990, the Hubble telescope has looked farther into space than anything before it, vastly aiding in our understanding of the universe.

HUBBLE'S SCOPE

Current
Universe

Normal
Galaxies

More Structured
Galaxies

Hubble
Deep Field

0

Distance (millions of light-years)

12,200

GRAVITATIONAL WAVES

Studying these space-time quakes, caused by extreme gravitational variations, allows for a better understanding of compact objects such as neutron stars and black holes.

ACROSS THE SPECTRUM

Current telescopes and observatories can capture the stars' entire electromagnetic spectrum, from visible light seen by telescopes like Hubble (which is also sensitive to near infrared and near ultraviolet) and radio waves detected by the Very Large Array (VLA), part of the National Radio Astronomy Observatory, to x-rays captured by the Chandra Observatory and the infrared perspective of the Spitzer Space Telescope. These images show the Crab Nebula over different wavelengths.

Radio Waves

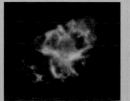

Infrared

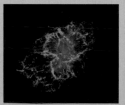

Visible Light

Ultraviolet

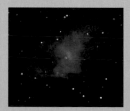

X-rays

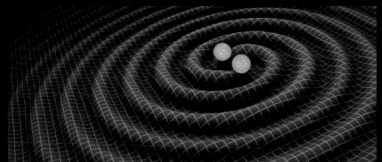

Hubble Ultra-Deep Field

First Galaxies

First Stars

Dark Age

Age of Radiation

Big Bang

13,000–13,300

13,800

AN ORDINARY

CALLED

STAR

THE SUN

THE SUN, THE FIRST OBSERVED STAR

The first explanations of stellar evolution used the sun as a reference point and were based on the principles of thermodynamics. These were limited, as they ignored nuclear fusion.

Long before we knew about nuclear fusion, scientists wondered what the mechanism was that gave the sun its power. Calculations done in the 19th century showed that if the source of the sun's energy was chemical, such as a combustion reaction, the sun would be able to sustain itself for only a few thousand years. Scientists like Hermann von Helmholtz claimed that the sun's brightness was due to compression caused by its own weight, or mass, which converted gravitational pressure or energy into heat—a mechanism governed by the laws of thermodynamics.

Radioactivity and Relativity
The discovery of radioactivity at the end of the 19th century and Albert Einstein's presentation in 1905 of the theory of special relativity, which linked space and time, energy and mass, made it possible to understand the true nature of the power of the sun and other stars.

The Road to Fusion
In 1920, Francis William Aston performed an experiment demonstrating that the mass of four hydrogen nuclei was greater than one helium nucleus, allowing astrophysicist Arthur Eddington to postulate that the sun's energy comes from the mass that remained during the conversion of hydrogen to helium. Today we know that nuclear fusion is what sustains a star's brightness over billions of years.

THE POWER OF THE SUN'S RADIATION
The sun's extraordinary power is especially intense for the planets closest to it in the solar system, to which the sun gives its name. Our planet's distance from this source of heat and light is what makes life possible.

WHAT IS THE SOLAR SYSTEM?
The solar system consists of the sun and the objects that orbit it: eight planets and their moons, as well as dwarf planets, asteroids, comets, dust belts, and tens of thousands of other objects.

THE EIGHT PLANETS
The eight planets in the solar system
[from nearest to the sun to farthest
away] are Mercury, Venus, Earth,
Mars, Jupiter, Saturn, Uranus, and
Neptune. Pluto was considered the
ninth planet until 2006, when it
was reclassified as a dwarf planet.
Mercury, Venus, and Earth are
shown here.

Stellar Evolution According to Thermodynamics

At the beginning of the 20th century,
astronomers Henry Norris Russell and
Norman Lockyer hypothesized that stars
develop when dispersed stellar material
clumps together, becoming increasingly
denser and hotter. A cold star (red) is
compacted until it reaches its maximum
brightness and temperature (blue), then it
cools and contracts until its death.

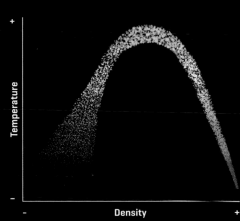

The Age of the Sun
In the mid-19th century, William Thomson
(who later became Lord Kelvin) proposed
another model to explain the increasing
density of stars, illustrated here: that
they are born blue or white and cool as
they contract. Basing his estimate on the
principles of heat transfer, he determined
that the sun was in an intermediate phase
of its evolution and calculated that it was 32
million years old, later declaring it up to 300
million years old. Scientists have estimated
the sun's age to be 4.6 billion years.

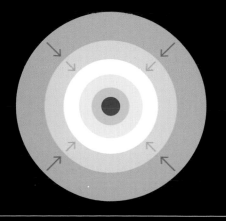

THE SUN IN THE MILKY WAY

The sun is in the center of our solar system, but not the Milky Way itself. It can be found orbiting 27,000 light-years away from the center and tens of thousands of light-years away from the outer rim, but it is constantly changing position, with the rest of the solar system in tow.

Before advanced telescopes, it was difficult to imagine that our solar system was part of something much larger: a galaxy with between 200 billion and 400 billion stars. Observing it is difficult because of the giant clouds of dust and gas that obscure some sections, but studying star clusters at the beginning of the 20th century helped us determine the position of the solar system. The sun takes between 225 million and 250 million years to orbit around the Milky Way's galactic center.

The Habitable Zone
The composition of the sun tells us where it first formed: near the galaxy's center. The first stars in the universe had only hydrogen, helium, and some other trace elements, but supernovae near the center of the galaxy sent out heavier elements, handing our solar nebula the raw materials needed for life. If it had been farther out in the galaxy, our solar system wouldn't have gotten the materials it needed to form as it did. The galactic habitable zone—the ring in which life can exist—extends between 13,000 and 32,000 light-years from the galactic center.

OUR POSITION IN THE GALAXY
The solar system is located in the Orion Arm of the Milky Way galaxy, in a zone where the ingredients necessary for life—in particular, water and carbon—are found.

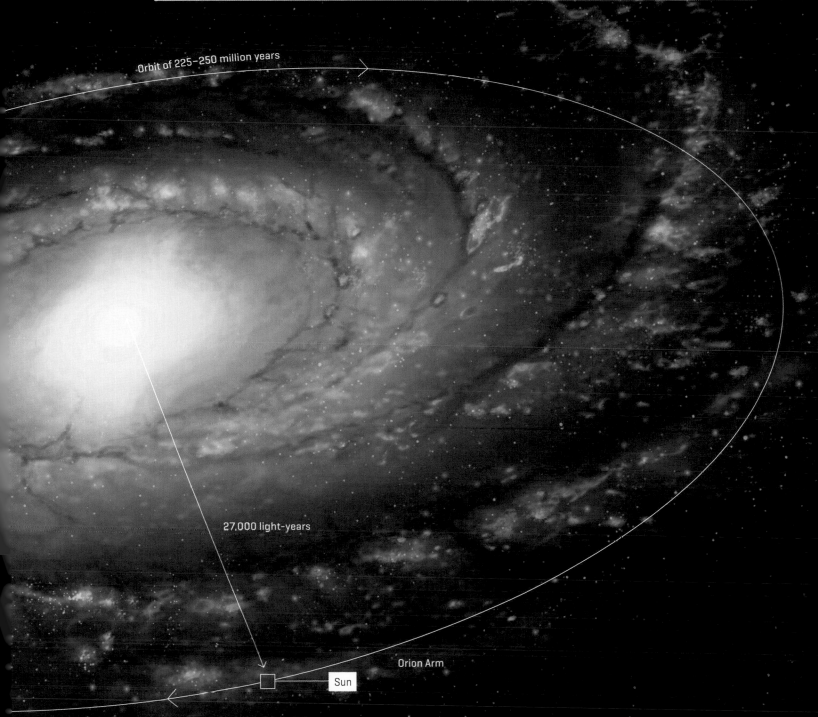

THE SPEED OF THE SUN

The sun moves around the galactic center at a speed of 225 to 250 kilometers (140 to 155 mi) per second, traveling in the direction of the Norma Cluster galaxies. This is fast when compared with the speed at which Earth orbits the sun, which is only 30 kilometers (18.6 mi) per second.

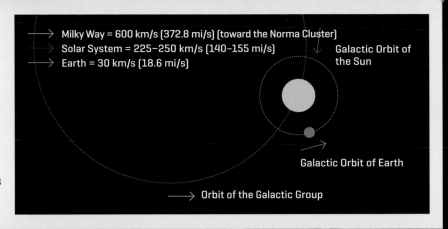

Milky Way = 600 km/s (372.8 mi/s) (toward the Norma Cluster)
Solar System = 225–250 km/s (140–155 mi/s)
Earth = 30 km/s (18.6 mi/s)

Galactic Orbit of the Sun

Galactic Orbit of Earth

Orbit of the Galactic Group

Orbit of 225–250 million years

27,000 light-years

Orion Arm

Sun

OUR GALACTIC ARM

From where we sit in the Milky Way, clouds of gas and dust make it impossible to view the galaxy as a whole. Even so, scientists have figured out where we are: in an area known as the Orion Arm because of its proximity to the constellation Orion.

Wezen

Eta Leonis

Betelgeuse

Orion's Belt

The Orion Arm (also known as the Local Arm) is located between the Sagittarius Arm and the Perseus Arm. It is some 3,500 light-years across and between 16,000 and 25,000 light-years long. The brightest stars we can see are located here. Sirius is the brightest star in the night sky, located some 8.6 light-years from Earth. Some of the farthest stars include Rigel, a system with two stars, which is between 700 and 900 light-years away. Our galactic neighborhood isn't very dense in comparison with the galactic center, but dense enough for some 2,000 stars to be found within a radius of 50 light-years.

The Gould Belt

The Orion Arm is home to a ringlike structure some 3,000 light-years across known as the Gould belt. It has a high concentration of young O- and B-type stars, including our sun. We don't know where the belt came from, although we think it was probably the result of a supernova or the collision of gas clouds in the galactic disc causing a wave of activity in the interstellar medium. This wave caused various molecular clouds to contract, creating neighboring stars.

THE MOST VISIBLE STARS
The Milky Way's two largest arms, the Sagittarius and Perseus Arms, flank the Orion Arm, represented in this illustration of some of the most prominent stars.

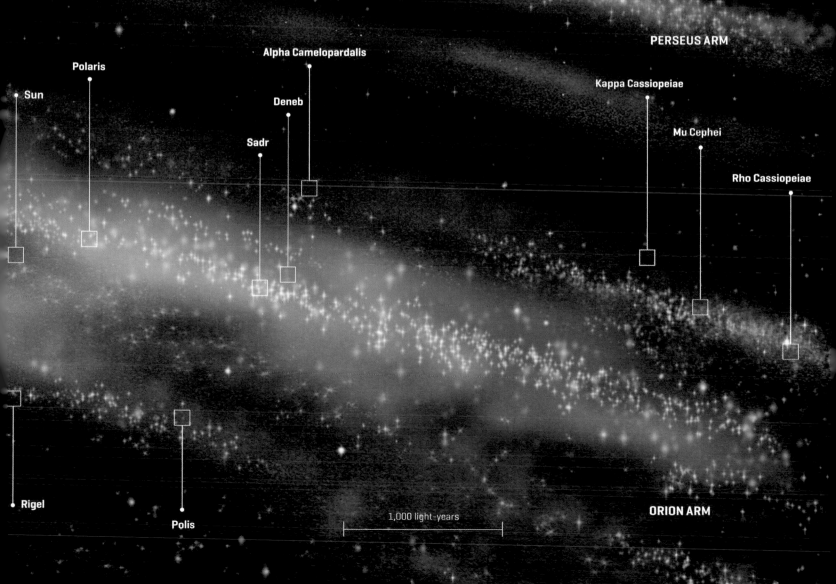

PERSEUS ARM

Alpha Camelopardalis

Polaris

Kappa Cassiopeiae

Sun

Deneb

Mu Cephei

Sadr

Rho Cassiopeiae

1,000 light-years

Rigel

Polis

ORION ARM

THE SUN'S NEIGHBORHOOD

Studying the stars nearest to our sun gives us an approximate idea of what other stars in the galaxy may be like.

In classifying stars using the Harvard classification system, only a few are B-, A-, or F-type (the hottest), while many are G- or K-type (mid-range temperatures), and several are M-type (the coolest). O-type stars, though rare, are the biggest and brightest, and have the shortest lives. They are not special except for the planets that orbit them, such as Ross 128 and Proxima Centauri B, two worlds whose temperatures are similar to Earth's. Some 14 planets have been identified less than 20 light-years from the solar system. It is believed that many stars have their own planets.

NAME OF THE STAR	DISTANCE TO THE SUN	NAME OF THE STAR	DISTANCE TO THE SUN
Proxima Centauri	4.24 light-years	Kruger 60 A, B	13.18 light-years
Alpha Centauri A, B	4.37 light-years	Ross 614 A, B	13.36 light-years
Barnard	5.96 light-years	Gliese 628 (Wolf 1061)	14 light-years
Wolf 359	7.78 light-years	Gliese 1	14.15 light-years
Lalande 21185	8.29 light-years	Wolf 424 A, B	14.30 light-years
Sirio A, B	8.58 light-years	Gliese 687	14.77 light-years
Gliese 65 A (Luyten 726-8A)	8.73 light-years	Gliese 674	14.81 light-years
UV Ceti (Luyten 726-8B)	8.73 light-years	GJ 1245 A, B, C	14.81 light-years
Ross 248 (HH Andromedae, Gliese 905)	10.30 light-years	Gliese 440 (LP 145-141)	15.11 light-years
Epsilon Eridani (Ran)	10.47 light-years	Gliese 876	15.20 light-years
Lacaille 9352	10.68 light-years	LHS 288	15.60 light-years
Ross 128	11.03 light-years	Gliese 1002	15.31 light-years
EZ Aquarii A, B, C	11.10 light-years	Gliese 412 A, B	15.81 light-years
61 Cygni A, B	11.41 light-years	Groombridge 1618	15.89 light-years
Procyon A, B	11.46 light-years	AD Leonis	16 light-years
Groombridge 34 A, B	11.62 light-years	Gliese 166 A, B, C	16.26 light-years
Epsilon Indi	11.81 light-years	Gliese 702 A, B	16.58 light-years
Tau Ceti	11.91 light-years	Altair	16.73 light-years
Gliese 1061	12.04 light-years	Gliese 570 A, B, C	19 light-years
YZ Ceti	12.20 light-years	Eta Cassiopeia A, B	19.42 light-years
Luyten's Star	12.20 light-years	Gliese 663 A, B	19.50 light-years
Teegarden's Star	12.52 light-years	Gliese 664	19.50 light-years
Kapteyn's Star	12.75 light-years	Gliese 783 A, B (HR 7703)	19.62 light-years
Lacaille 8760	12.87 light-years	Delta Pavonis	19.92 light-years

LOCATIONS OF THE CLOSEST STARS
This diagram includes stars that are less than 20 light-years from the sun. The colors of the stars correspond to their temperature: blue (class O) is the hottest and red (class M) is the coldest.

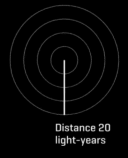

Distance 20 light-years

B C

Gliese 166

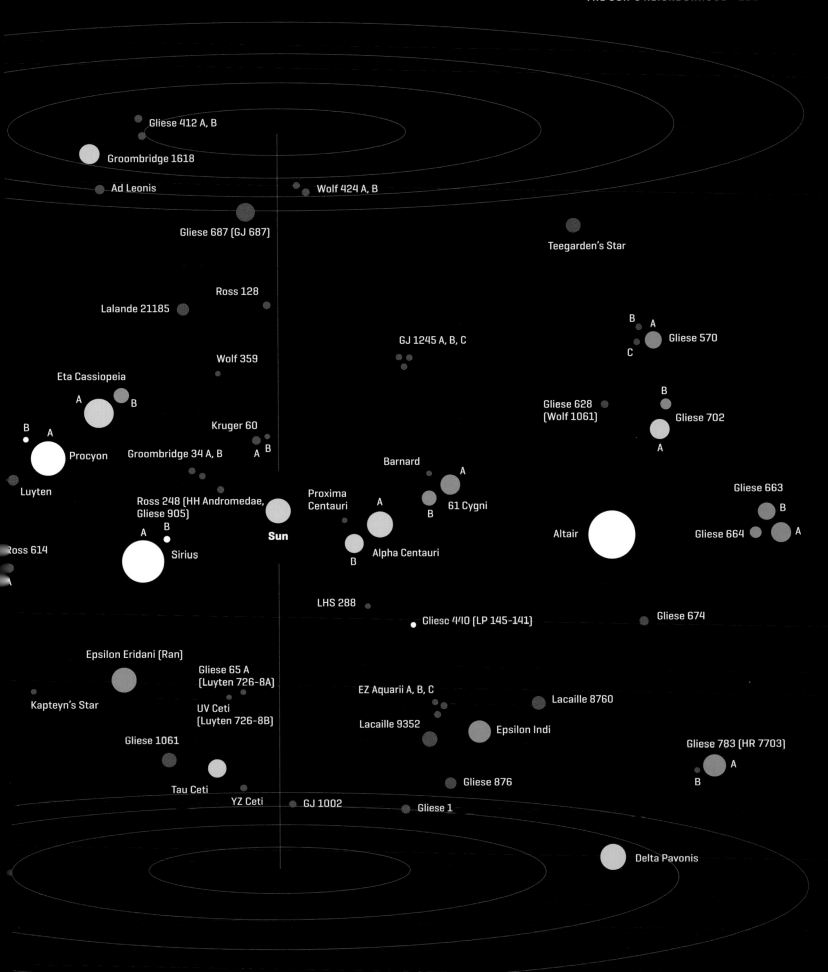

Gliese 412 A, B

Groombridge 1618

Ad Leonis

Wolf 424 A, B

Gliese 687 (GJ 687)

Teegarden's Star

Ross 128

Lalande 21185

GJ 1245 A, B, C

Wolf 359

B   A
Gliese 570
C

Eta Cassiopeia

A   B

Gliese 628
(Wolf 1061)

B
Gliese 702

A

B   A

Procyon

Kruger 60

Groombridge 34 A, B   A   B

Barnard

A

Luyten

B

61 Cygni

Gliese 663

B

Ross 248 (HH Andromedae,
Gliese 905)

Proxima
Centauri

A

Gliese 664   A

A   B

Altair

Ross 614

Sirius

Sun

Alpha Centauri

B

LHS 288

Gliese 674

Gliese 440 (LP 145-141)

Epsilon Eridani (Ran)

Gliese 65 A
(Luyten 726-8A)

EZ Aquarii A, B, C

Lacaille 8760

Kapteyn's Star

UV Ceti
(Luyten 726-8B)

Lacaille 9352

Epsilon Indi

Gliese 1061

Gliese 783 (HR 7703)

A

B

Tau Ceti

Gliese 876

YZ Ceti   GJ 1002   Gliese 1

Delta Pavonis

THE SUN'S SIZE

In galactic terms, our sun is not particularly special, but it is the largest object in the solar system—massive compared to everything around it—and produces an extraordinary amount of energy.

The sun is a very normal star, galactically speaking. Its mass is not particularly noteworthy—in fact, it is actually quite modest in comparison with the universe's most massive stars. But it is the largest object in the solar system, containing 99.86 percent of its total mass. The gravity of its photosphere, the outer zone of the sun's atmosphere from which light radiates, is 28 times that of Earth's. Its radius is more than 109 times that of Earth's, its volume more than 1,304,000 times. However, its rotation is much slower than our planet's and changes depending on its latitude. Its equator takes approximately 26 days to do a full rotation, while the poles need more than 30 days.

Star Statistics

The exact composition of the sun is still up for debate, but we know that it is formed mostly of hydrogen and, because of hydrogen fusion, helium. The amount of energy this G-type star produces per second could supply the human race at its current energy consumption rate for the next 675,000 years.

Average rotation	27 days, 6.5 hours
Diameter	1,392,000 km
Mass	1.9891×10^{30} kg
Volume	1.4123×10^{18} km^3
Gravity on the surface	274 m/s^2
Density	1,411 kg/m^3
Luminosity	$3,827 \times 10^{26}$ W
Apparent magnitude	-26.8
Core temperature	15,500,000°C
Surface temperature	5500°C

The Sun's Makeup

The sun is composed mostly of hydrogen and a product of its nuclear fusion, helium, but there are also small amounts of other elements. Seventy of them have been detected so far.

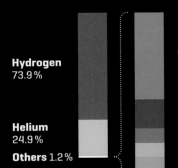

Hydrogen 73.9%

Helium 24.9%

Others 1.2%

Oxygen 0.6%

Carbon 0.2%

Iron 0.1%

Silicon 0.1%

Nitrogen 0.1%

Neon 0.05%

Magnesium 0.05%

AN IMMENSE STAR

This photograph shows a section of the sun as seen from space. The tiny granular details barely seen on the immense surface of the sun are, in fact, giant features.

SUNLIGHT'S TRAVELS

The energy generated in the sun's core takes thousands of years to reach its surface. A photon (a particle that carries energy in the form of light) travels along a winding route, colliding with other particles until it is able to break free from the sun's atmosphere and travel through the solar system. This illustration shows approximately how many minutes it takes for that light to reach each planet. The distances are not to scale.

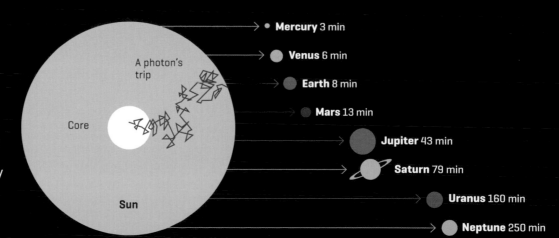

Core

A photon's trip

Sun

Mercury 3 min

Venus 6 min

Earth 8 min

Mars 13 min

Jupiter 43 min

Saturn 79 min

Uranus 160 min

Neptune 250 min

A POWERHOUSE OF ENERGY

The sun is an enormous nuclear reactor in which energy, produced in the core through nuclear fusion, flows through its layers and into its atmosphere, where it is finally released into space.

The temperature (some 15.5 million degrees Celsius) and density in the sun's core are so extreme that hydrogen atom nuclei are forced to combine, creating helium atom nuclei. This fusion process, called a proton–proton chain, produces huge amounts of energy that is transported by photons made in the sun's core and escapes through the radiative layer, a high-density region with a temperature that varies between 1.5 million and 15 million degrees Celsius. The opacity of this layer (its impenetrability to electromagnetic radiation, which includes photons) becomes so high that photons are no longer able to pass through it and transport energy effectively. Hot gases, called plasma, bubble up and form a convection system, where the plasma transfers its energy, cools down, and then sinks again, pushing energy toward the sun's surface.

Release of Energy

After reaching the upper part of the convection layer, photons can escape into space through the photosphere, the surface of the sun (which can be seen from Earth). Above that is the chromosphere, in which the temperature increases from 5500 to over 20,000 degrees Celsius (9932–36,032°F). The last layer of the sun's atmosphere is the corona, made of very diluted, yet extremely hot, plasma. Why the corona is so hot is still unknown.

This photograph shows the sun's photosphere, the only layer that can be observed with the naked eye.

The Sun's Layers

Chromosphere
The high temperatures in the chromosphere cause hydrogen to emit a reddish light. That light gives this layer of the sun its name (chromo = color).

Core
This is where the main nuclear fusion reactions are produced, thanks to enormous temperatures that range from tens of millions to billions of degrees.

Convection Layer
In this layer, hot plasma (gases that conduct electrical currents) rises, transfers its energy, cools, and sinks again in a circular motion known as convection. The temperature at the bottom of the convection layer is 200,000 degrees Celsius (360,032°F). At the top of the convection layer (near the surface of the sun), it is only about 5700 degrees Celsius.

Corona
The outermost part of the sun's atmosphere can reach temperatures in the millions of degrees. It appears as a whitish halo during a total solar eclipse.

Photosphere
The visible surface of the sun, which emits the radiation, or visible sunlight, that reaches Earth. Because the sun is a ball of gases, there is no solid surface.

Radiative Layer
The radiative layer accounts for about 70 percent of the sun's radius. The density of this layer is so great that some of the core's energy in the form of photons (light particles) is absorbed here.

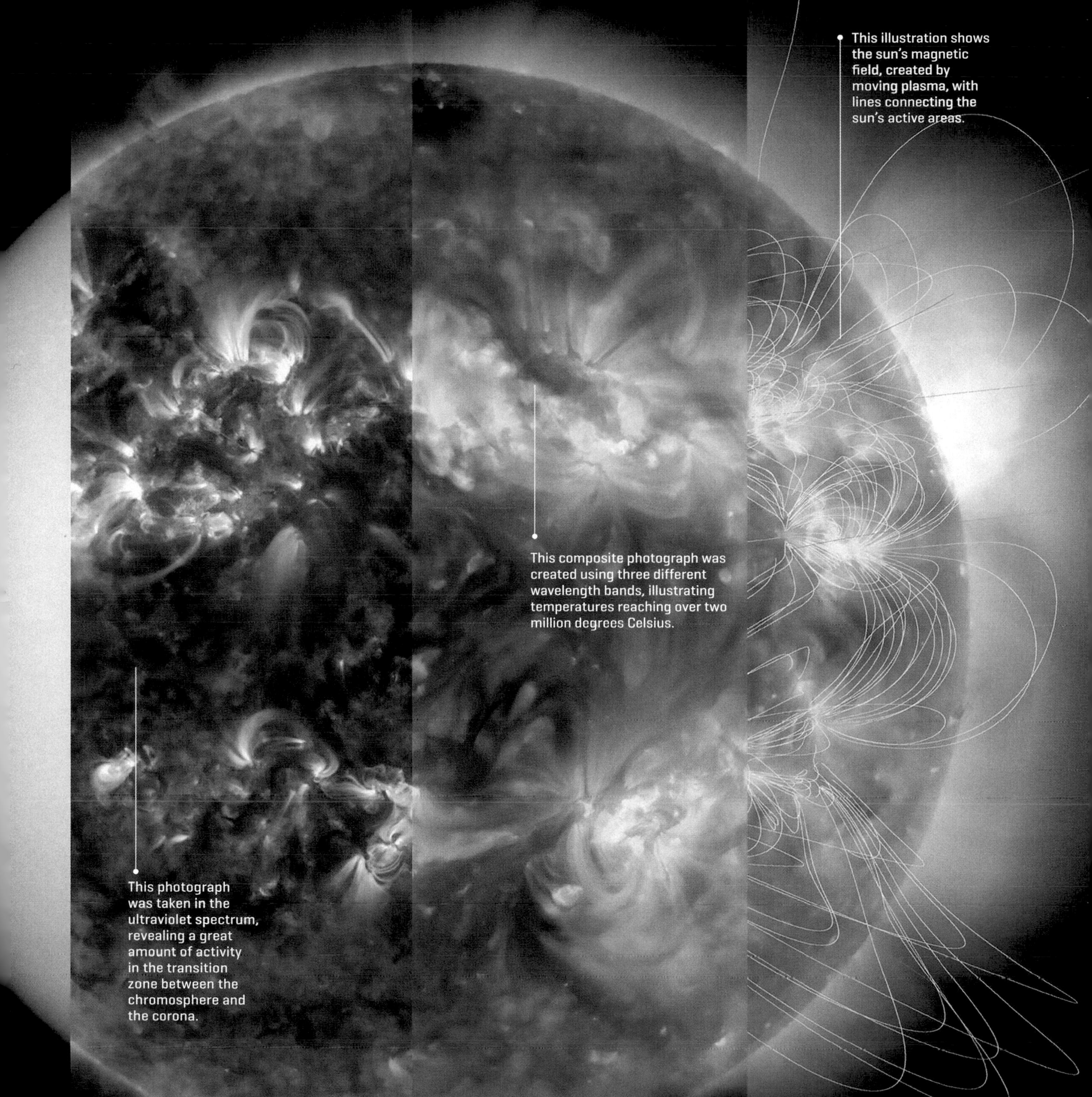

This illustration shows the sun's magnetic field, created by moving plasma, with lines connecting the sun's active areas.

This composite photograph was created using three different wavelength bands, illustrating temperatures reaching over two million degrees Celsius.

This photograph was taken in the ultraviolet spectrum, revealing a great amount of activity in the transition zone between the chromosphere and the corona.

DIFFERENCES IN THE LAYERS OF THE SUN

This composite image uses three photographs taken at different wavelengths, along with a complementary illustration that shows the sun's magnetic field.

THE SOLAR CYCLE

The sun is composed of electrically charged gases, or plasma. These gases move, creating a powerful magnetic field. Periodic changes in the magnetic field, as well as changes in the sun's appearance in the form of sunspots and flares, are all part of the solar cycle, also called the solar magnetic activity cycle.

About every 11 years the sun's magnetic field inverts, or flips: the north pole becomes the south and the south becomes the north. It takes another 11 years or so for the poles to flip back. As the magnetic field changes, so does the level of activity, creating sunspots, flares (or solar eruptions), and giant explosions known as coronal mass ejections. We track the solar cycle by counting its sunspots. The beginning of a solar cycle is called the solar minimum, which is when the sun has the fewest sunspots. Over time solar activity increases and the number of sunspots grows. The middle of the solar cycle, the solar maximum, is when the sun sees the most activity, and thus the most sunspots. As the cycle ends, the sun goes back to the solar minimum and a new cycle begins.

What Are Sunspots?

Sunspots are dark, cooler spots on the surface of the sun. These areas are cooler than the rest because they are where the sun's magnetic field is strongest, keeping heat from the core from reaching the surface.

Solar Minimum
During the solar minimum, the corona is at its calmest state, with few areas of notable activity.

Flipping the Poles

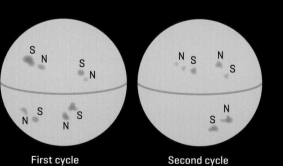

First cycle
of 11 years

Second cycle
of 11 years

During the solar maximum phase, the sun's magnetic poles invert: the north pole becomes the south pole and vice versa. They do this every 11 years or so.

SOLAR IRRADIANCE

Solar irradiance, or the energy received per unit of surface and temperature, is constant on Earth, although this value changes depending on solar activity. As can be seen in the figure, there is a close correlation between the number of sunspots and solar irradiance.

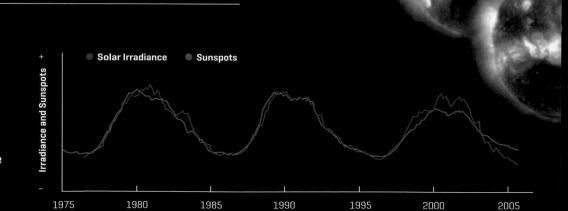

Irradiance and Sunspots

● Solar Irradiance   ● Sunspots

1975   1980   1985   1990   1995   2000   2005

1999
2000
2001
2002
2003

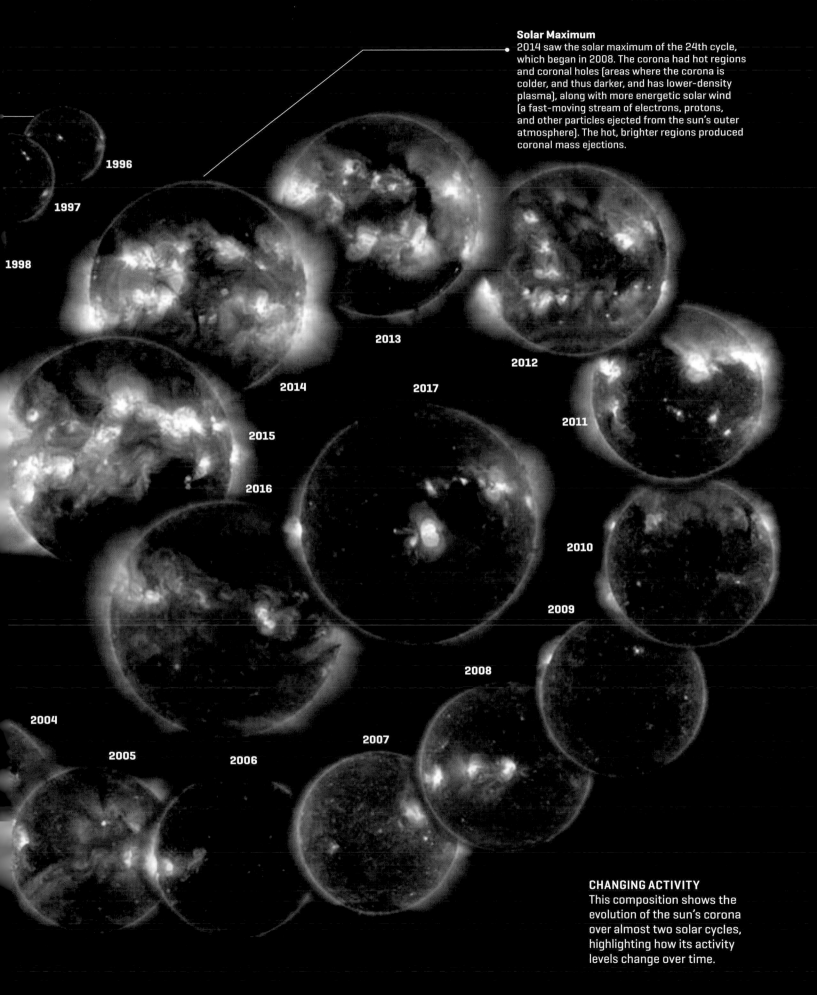

Solar Maximum
2014 saw the solar maximum of the 24th cycle, which began in 2008. The corona had hot regions and coronal holes (areas where the corona is colder, and thus darker, and has lower-density plasma), along with more energetic solar wind (a fast-moving stream of electrons, protons, and other particles ejected from the sun's outer atmosphere). The hot, brighter regions produced coronal mass ejections.

1996

1997

1998

2013

2012

2014

2017

2011

2015

2016

2010

2009

2008

2004

2007

2005

2006

CHANGING ACTIVITY
This composition shows the evolution of the sun's corona over almost two solar cycles, highlighting how its activity levels change over time.

THE SUN'S ATMOSPHERE

The sun's atmosphere, made up of the photosphere, chromosphere, and corona, releases energy into space. The temperature of these layers ranges from several thousand to millions of degrees.

The atmosphere's outermost layer is the photosphere, which can be seen from Earth and gives the sun its characteristic color. The photosphere is in contact with the convection layer, but it allows photons to escape into space.

A Violent Connection with Space

The second layer is the chromosphere, which separates the photosphere and the outermost layer of the sun's atmosphere. It is more than 1,000 kilometers (621 mi) wide and produces some of the sun's most violent events, such as solar flares and prominences. These events cause ejections of immense clouds of high-energy plasma that can be harmful to spacecraft and satellites that orbit Earth. The outermost region of the sun, the corona, contains extremely high-temperature gases.

Atmosphere Densities

This representation of the sun's atmospheric layers shows the change in density in relation to the altitude above the photosphere, or surface.

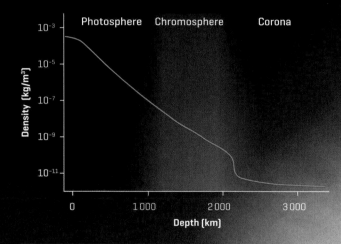

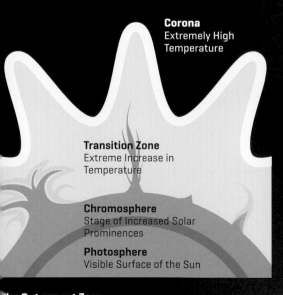

Corona
Extremely High Temperature

Transition Zone
Extreme Increase in Temperature

Chromosphere
Stage of Increased Solar Prominences

Photosphere
Visible Surface of the Sun

The Outermost Zone
Violent phenomena occur in the outermost zone of the sun that are notable for their complexity, but continue to defy our understanding.

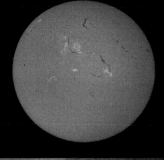

PHOTOSPHERE
In this region, plasma's opacity decreases enough to allow photons to escape into space, causing sunspots and faculae, or solar bright spots.

CHROMOSPHERE
This layer is where solar flares (also known as eruptions) and prominences (large, gaseous loops) occur.

CORONA
Although the temperature is very high in this outermost region, its density is low enough to make it hard to see. Plasma escapes it as solar wind.

VIOLENT SOLAR FLARES
Disruptions in the sun's magnetic field cause powerful flares in the sun's atmosphere.

S☉LAR LIGHT

The sun's light, both visible to us and not, contains rich information about its physical and chemical characteristics.

The solar spectrum is like a rainbow but with many more shades, including both visible light and electromagnetic radiation. The light spectrum that reaches Earth is so strong it can even be measured with outdated instruments. Interactions between solar light and stellar material leave traces that can be detected in the spectrum as dark bands as atoms of each element in the solar atmosphere tend to absorb light at specific wavelengths. The absence of light in the spectrum causes these dark bands, known as absorption lines, which can be used to identify the elements that cause them.

Telltale Lines

The absorption lines of iron, sodium, magnesium, calcium, and other elements show that the sun was created from the remnants of earlier stars, as those elements did not exist before the first stars.

Electromagnetic Spectrum

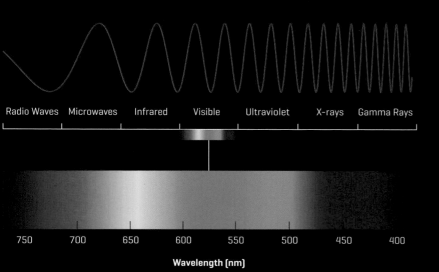

| Radio Waves | Microwaves | Infrared | Visible | Ultraviolet | X-rays | Gamma Rays |

750   700   650   600   550   500   450   400

Wavelength (nm)

Light consists of photons: electromagnetic waves with different wavelengths. Those with higher wavelengths are radio waves, which have no upper limit. The rest of the bands follow in gamma rays are the shortest at a billionth of a meter or less. Visible light is a minuscule part of the electromagnetic spectrum—each color only covers some tens of millionths of a millimeter.

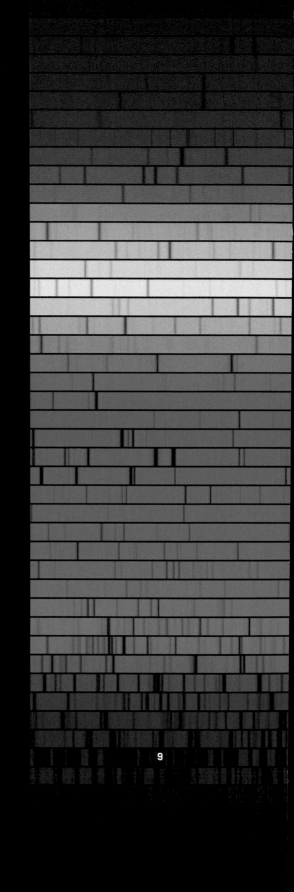

9

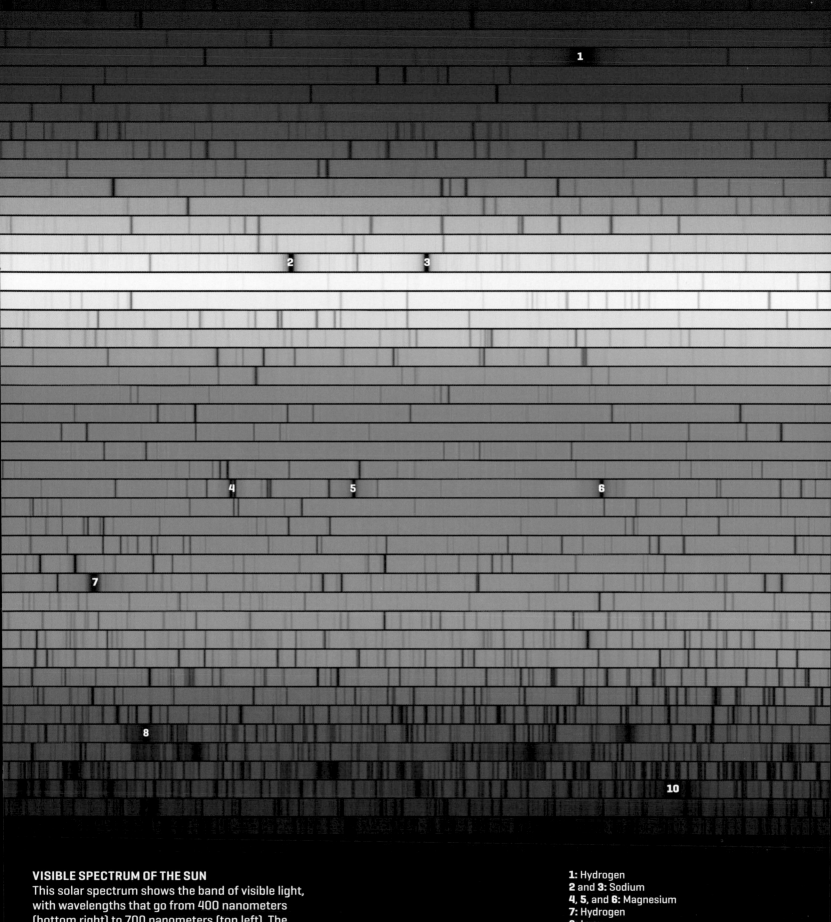

VISIBLE SPECTRUM OF THE SUN
This solar spectrum shows the band of visible light, with wavelengths that go from 400 nanometers (bottom right) to 700 nanometers (top left). The numerous absorption lines betray the presence of chemical elements.

1: Hydrogen
2 and **3:** Sodium
4, 5, and **6:** Magnesium
7: Hydrogen
8: Iron
9: Iron (a group of very close lines)
10: Calcium

THE CREATION OF THE SUN

Some 4.6 billion years ago, a nearby supernova disrupted a molecular cloud, or it passed through a density wave of a galactic arm. This disrupted the cloud's equilibrium, causing it to collapse and, eventually, to give birth to a group of stars, including the sun.

Molecular clouds, sometimes called stellar nurseries, are immense accumulations of very low temperature gas and dust. They are huge, reaching several millions of solar masses, and their diameter can be hundreds of light-years. Inside these clouds, high-density clumps of stellar material begin to contract and eventually break into several dense regions, each of which becomes a star. The result is an open stellar cluster that can contain tens to several thousands of stars. In the distant past, our sun was part of a cluster, but its stars drifted apart over time. Astrophysicists are searching for the sun's sister stars, lost long ago in the vast expanse of our galaxy.

A PROCESS DIRECTED BY GRAVITY

Action of the Force of Gravity

The Primordial Cloud
The sun formed inside a large cloud of gas (mostly hydrogen, with some helium) and dust, the likes of which can be found in other parts of the cosmos.

The Trigger
Some 4.6 billion years ago, a part of the molecular cloud began to contract, perhaps due to a shock wave from a nearby supernova. Gravity continued this process.

Internal Differentiation
The cloud did not contract into just one ball of material, but several. These clumps in turn formed other, larger, and more cohesive groups.

Star-forming Seeds
A compressed bit of cloud fractured into several star-forming seeds, which continued to evolve independently. One of these became our sun.

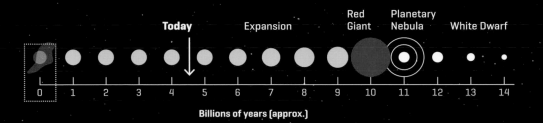

Today | Expansion | Red Giant | Planetary Nebula | White Dwarf

0 1 2 3 4 5 6 7 8 9 10 11 12 13 14

Billions of years (approx.)

LIFE CYCLE

Since its initial formation, the sun has moved into a long and relatively stable phase of its evolution. It will continue along this path, eventually becoming a red giant before turning into a planetary nebula and a white dwarf.

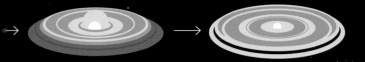

Embryo
The final contraction of a star-forming seed created an early form of the sun, known as the protosun, surrounded by a spinning disc of material. Its gravitational pull drew in 99.9 percent of the mass of the future solar system.

Birth
As it gained mass, the protosun eventually reached a critical pressure and temperature point to begin fusing hydrogen atoms and creating helium nuclei, releasing a large amount of energy.

PROTOPLANETARY DISC OF OUR STAR
In this illustration, the protosun is surrounded by a large amount of dust and gas ready to condense into planetesimals: the seeds of future planets and other bodies in the solar system.

LEAVING THE STELLAR NURSERY
Seen here with visible light, the young stars of the Eagle Nebula are leaving it slowly, just as the sun left its molecular cloud. The gaseous cocoon in which they formed is being swept away by ultraviolet radiation.

MATURATION OF THE SUN

In its infancy, the sun maintained its surrounding disc of material, which went on to form other bodies. Its disc dissipated by the time the sun entered middle age, surrounded instead by various planets.

It was 30 percent less bright then than it is today and, according to theories of physics, its brightness should continue increasing as it expands. Though it's no longer a young star, it hasn't started to deteriorate with old age. When it does, it will experience violent structural changes and won't be able to emit energy like it once did.

A Stable Life-Giving Agent

The sun has been stable for quite a long time. That's a very positive thing for life on Earth, which has enjoyed a long, peaceful period with perfect conditions for species like humans to evolve. It will stay stable for several million more years, but this peace definitely won't last forever.

ORIGIN OF WORLDS
Seen in profile, this bright disc of matter around the sun formed the planets.

PLANET EMBRYOS
Material from the sun's disc formed planetesimals, the building blocks for the construction of planets.

Today | Expansion | Red Giant | Planetary Nebula | White Dwarf

0 1 2 3 4 5 6 7 8 9 10 11 12 13 14

Billions of years (approx.)

SOLAR "WAVE"

Just as seismologists learn about the interior of our planet by studying Earth's tremors, solar physicists and helioseismologists can learn about the interior of the sun in the same way. They analyze detectable solar vibrations, or "waves," on the star's photosphere, then create computer models such as the one shown on the right. These waves propagate about 100 times faster than sound does through air.

DEFINED IDENTITY
Before the sun became an adult, each planetesimal already had a distinct planetary shape.

A NEW WORLD
As the Earth finished forming, the sun entered its long middle age of hydrostatic equilibrium, burning steadily and giving us the perfect conditions for life.

THE SUN AS A RED GIANT

When the hydrogen in the sun's core runs out, so will the pressure that maintains its equilibrium. The core will begin to contract, releasing gravitational energy that will signal the beginning of the star's end.

Some 10 billion years after its birth, the sun's core will start to run out of hydrogen, leaving mostly helium that the star isn't hot enough to burn. Without hydrogen fusion to push back against gravity, the sun will collapse inward, increasing in density and temperature until it begins fusing the hydrogen surrounding its core. It will then expand and cool, becoming a red giant. This renewed activity will heat its helium core, and once it reaches some 100 million degrees Celsius it will begin to produce carbon. The sun will continue fusing helium for 100 million years, and fusing hydrogen in its outer layer for much longer. Its hydrogen-burning shell will create an outward push, causing the star to expand.

The Evolution of the Habitable Zone

The zone around the sun in which water can exist as a liquid includes Earth and extends to Mars, but it won't stay there. When the sun becomes a red giant, swallowing both Mercury and Venus, the habitable zone will shift from Mars to as far out as Saturn. This image, which is not to scale, shows its projected future.

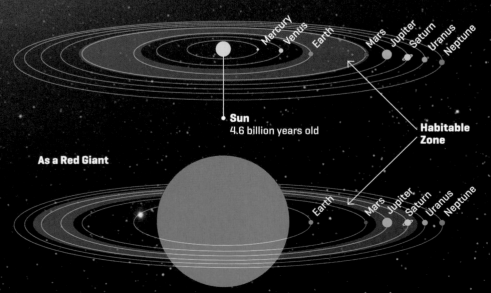

Today

Mercury · Venus · Earth · Mars · Jupiter · Saturn · Uranus · Neptune

Sun
4.6 billion years old

Habitable Zone

As a Red Giant

Earth · Mars · Jupiter · Saturn · Uranus · Neptune

Sun
10 billion years old

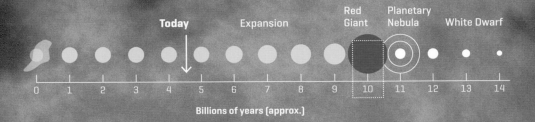

Today   Expansion   Red Giant   Planetary Nebula   White Dwarf

0 1 2 3 4 5 6 7 8 9 10 11 12 13 14

Billions of years (approx.)

THE FINAL FRONTIER
A red giant can grow large
enough to devour the
closest orbiting planets,
as the sun will do to
Mercury and Venus.

As a Red Giant
The sun's core will
contract as its shell
expands, creating a red
giant with a degenerate
carbon oxygen core.

DEATH OF THE SUN

The mass of the sun will determine its end. Since it's a low-mass
main-sequence star, when it runs out of fuel it will become a
planetary nebula around what will become a white dwarf.

The nuclear process by which the sun
gets its energy involves the irreversible
transformation of four protons into a
helium-4 nucleus. Within five billion
years, it will run out of hydrogen and
its central region will be made up of
helium. The proton–proton chains will
disappear, causing its core to begin
collapsing inward. It will heat up as it
collapses, and when the temperature
reaches 100 million degrees Celsius
then helium fusion will begin, creating
carbon-12 and oxygen-6. The sun

fusion reactions for a while, but when
the helium runs out it will have no
source of energy, as its mass won't be
great enough to fuse carbon.

A Core Without a Shell
With two levels of hydrogen and
helium combustion, an intense solar
wind will rip off the sun's shell, leaving
its carbon-oxygen core behind. That
core will become a very young white
dwarf, while its shell will become a
beautiful planetary nebula.

Planetary Nebula
A period of intense solar
wind will eject the sun's
shell, which will become a
planetary nebula, while its
naked core will become a
proto-white dwarf.

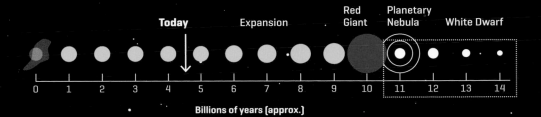

Today Expansion Red Giant Planetary Nebula White Dwarf

0   1   2   3   4   5   6   7   8   9   10   11   12   13   14

Billions of years (approx.)

PATH TO THE END
The sun's final-life stage will start when it becomes a red giant and finish when it shrugs off its external layers, its remaining core turning into a faint white dwarf.

A Naked Core
When the shell is ejected the sun will be reduced to its central core, whose surface temperature will be around 100,000 degrees Celsius.

The End: White Dwarf
Over billions of years, the core's temperature will decrease and the white dwarf, having lost most of its thermal energy, will become nearly invisible.

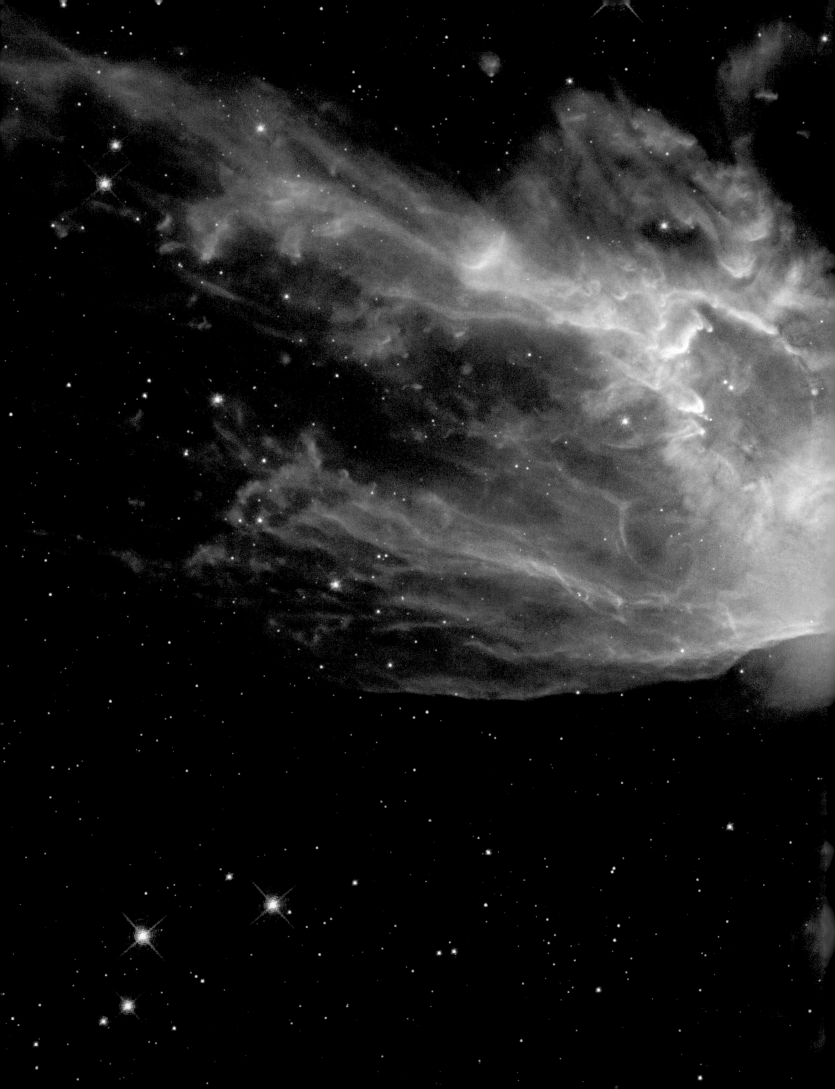

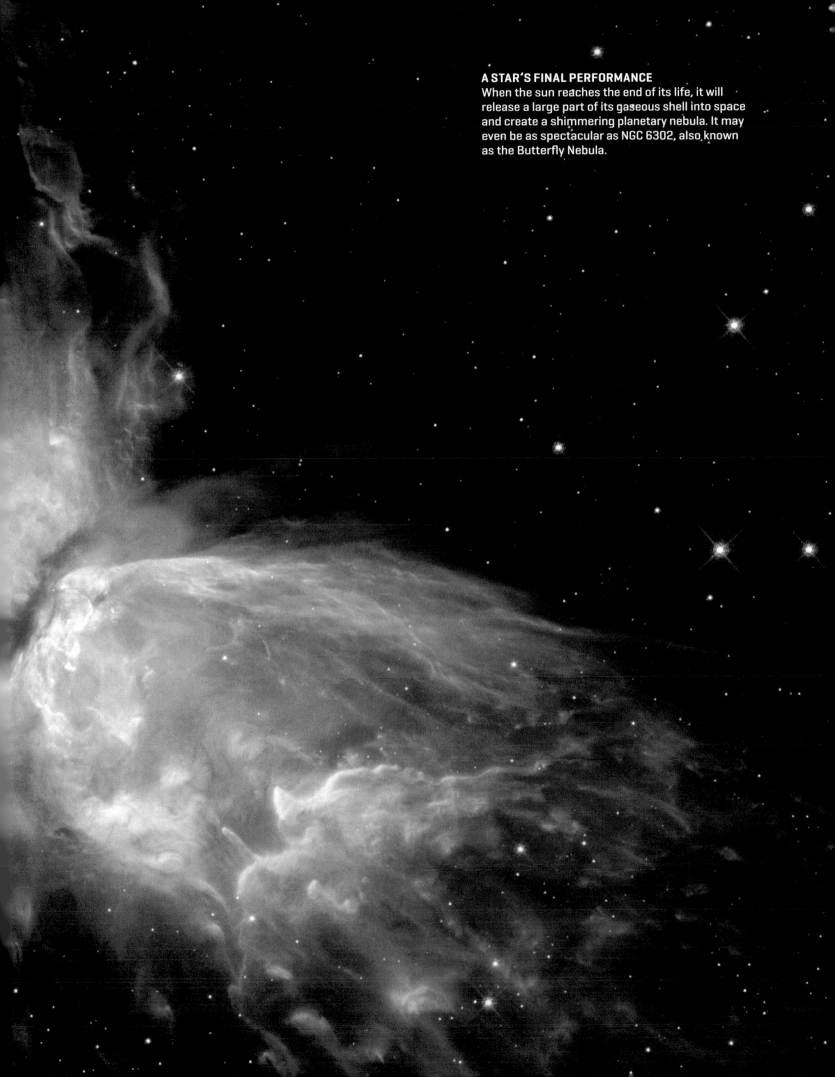

A STAR'S FINAL PERFORMANCE
When the sun reaches the end of its life, it will release a large part of its gaseous shell into space and create a shimmering planetary nebula. It may even be as spectacular as NGC 6302, also known as the Butterfly Nebula.

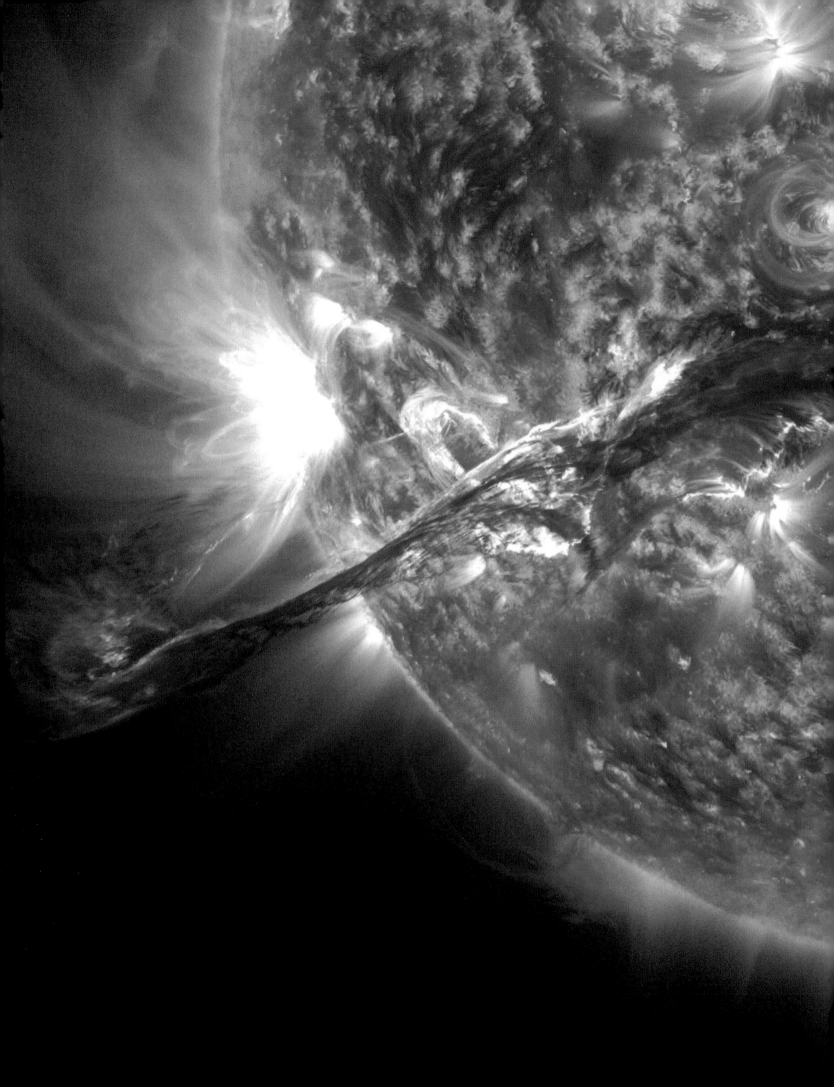

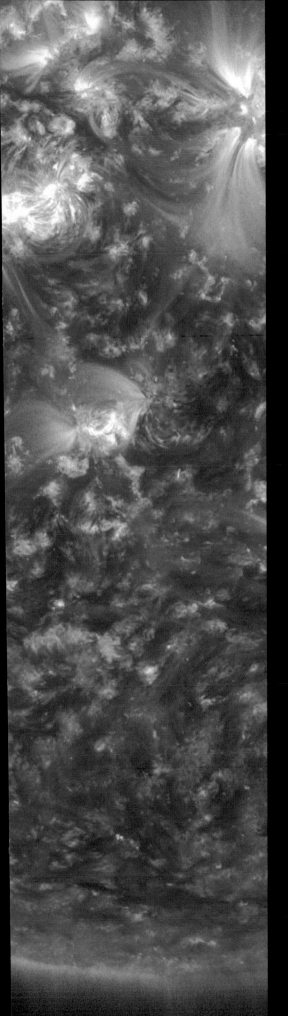

THE SUN AND ITS INFLUENCE ON EARTH

The sun bursts to life with a solar eruption, a phenomenon that occurs during its most active periods. Coronal mass ejections release so much energy that they can disrupt electrical networks, satellites, and other technological

THE SUN'S MAGNETIC FIELD

The sun's magnetic field affects solar activity such as sunspots and flares, but it also affects the planets in the solar system.

The sun's solar wind carries its magnetic field into the heliosphere, the region surrounding the sun and the planets. This interplanetary, or heliospheric, magnetic field has a spiral shape, created by the sun's rotation (the sun completes one rotation about every 27 days).

Dark Spots

The sun's magnetic field stops convection, reducing its efficiency in terms of heat transport. Where the field is the most intense, dark sunspots appear. These spots are approximately 1,300K colder than the rest of the photosphere, which is why they appear darker than the plasma surrounding them.

The Magnetic Field's Spiral

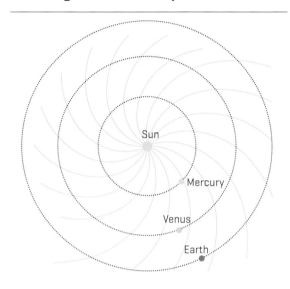

Solar wind, which is made of protons, electrons, and other high-energy particles, carries the magnetic field outward through the solar system. This interplanetary magnetic field moves out in a spiral pattern that fills the space between the sun and its planets.

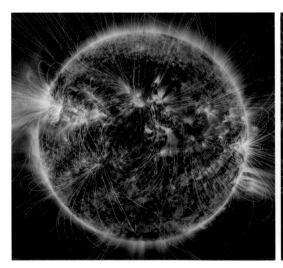
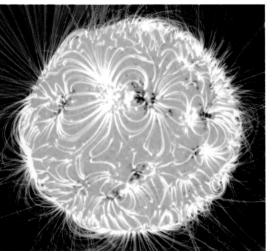

VIEWING THE SUN'S MAGNETIC FIELD
In the image on the left, magnetic field lines have been added to a photograph obtained by NASA's Solar Dynamics Observatory (SDO). On the right is a computer model of the sun's magnetic field.

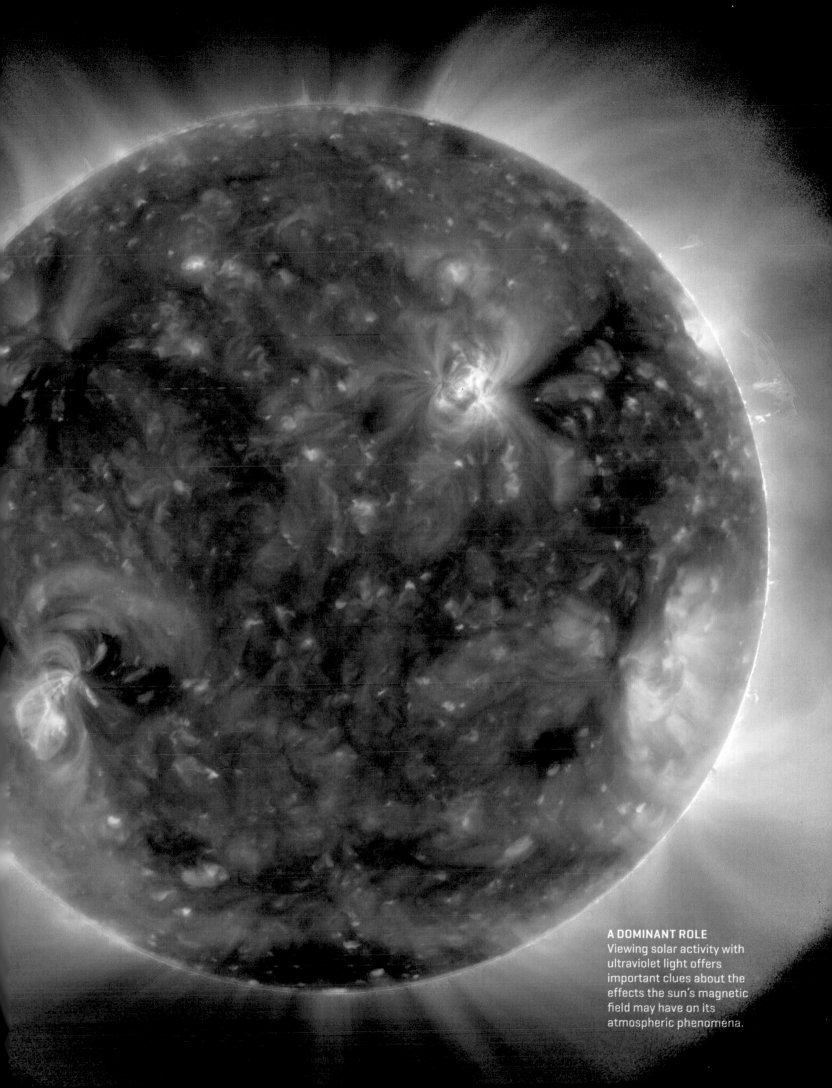

A DOMINANT ROLE
Viewing solar activity with ultraviolet light offers important clues about the effects the sun's magnetic field may have on its atmospheric phenomena.

SOLAR WIND

As the solar wind leaves the sun's corona, it carries the sun's magnetic field lines into space, forming the interplanetary magnetic field. The solar wind moves at speeds from 300 to 1,000 kilometers per second (186–621 mi/s). The sun has lost 0.01 percent of its mass due to this influential outflow of electrons, protons, and other charged particles.

Studying the solar wind helps scientists understand the composition of the sun and the magnetic field. Data for analysis have come from experiments on the moon conducted during the Apollo program and through NASA's Genesis probe.

Effects of the Solar Wind

Earth, like most other astronomical bodies, has its own magnetic field, called the magnetosphere. The magnetosphere acts as a protective bubble, deflecting the solar wind and reducing the number of solar electrical particles that penetrate Earth's atmosphere. Mars does not have a protective magnetosphere, and thus has lost much of its atmosphere.

The Magnetic Field's Changing Speeds

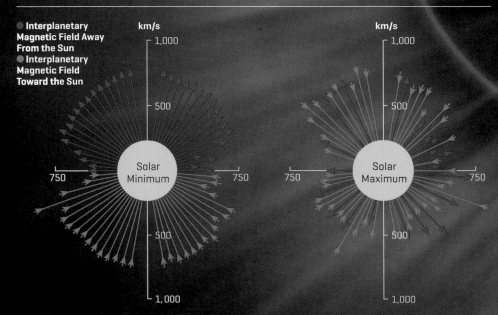

● Interplanetary Magnetic Field Away From the Sun
● Interplanetary Magnetic Field Toward the Sun

km/s
1,000
500
Solar Minimum
750   750
500
1,000

km/s
1,000
500
Solar Maximum
750   750
500
1,000

The Ulysses probe observed our sun at a wide range of latitudes, revealing new information about its heliosphere. The length of the lines indicate speed, while the color indicates the direction of the interplanetary magnetic field: red lines are moving away from the sun and blue are moving toward it. During a solar minimum, solar wind speed is lower along the sun's equator and higher at its polar regions, and the interplanetary magnetic field is bipolar. During a solar maximum, the magnetic field structure is much more complex, with fast and slow solar winds found at all latitudes.

Heliosphere (0–121 AU)
The solar wind moves within the sun's area of influence.

Termination Shock (80–100 AU)
The point at which the solar wind speed is reduced to practically zero.

Interstellar Medium (121 AU)
Outside the solar system, the interstellar medium is made of the gas and dust found between star systems.

Sun
Every year the sun loses some 2 × 10^{-14} solar masses of particles carried away by solar wind.

Neptune (30 AU)
The farthest planet from the sun, and the last one impacted by its solar wind.

Bow Shock (230 AU)
This outward ring around the heliosphere is the bow shock, created by the collision of the solar wind and the interstellar medium.

Heliopause (121 AU)
This is the changing front between the heliosphere and the interstellar medium, which lessens the solar wind's strength.

THE SOLAR WIND'S REACH

The solar wind extends from the heliosphere to well beyond Neptune's orbit. Its effects ebb at the heliopause, where the interstellar medium diminishes the wind's strength. The termination shock is the point at which the solar wind slows abruptly. The bow shock, named for the wave made by a ship's bow as it moves through water, is created where two streams of gas collide: in the sun's case, solar wind and the interstellar medium. The distances shown above are measured in astronomical units (AU).

ABOVE NORWAY
From Earth's surface, this aurora looks like a ghostly, semitransparent curtain.

AN AURORA FROM SPACE
Aurorae can be observed clearly from the International Space Station, revealing more detail than can be seen from Earth.

IN ALASKA
This image obtained from Eielson Air Force Base shows the intricate nature of an aurora.

AURORAE

Shimmering polar aurorae are some of the most spectacular and beautiful phenomena in nature. They are caused by the clash between the sun's electrically charged particles, carried by the solar wind, and Earth's upper atmosphere.

Earth's magnetosphere keeps out most of the sun's electrically charged particles, but some slip through. When they do, they are attracted to the magnetic poles, where they collide with chemicals in our atmosphere to create beautiful light shows, or aurorae: the aurora borealis in the Northern Hemisphere and aurora australis in the Southern Hemisphere. They are only visible at latitudes below 50 degrees under exceptional conditions, such as a coronal mass ejection.

Creating the Lights
When solar particles collide with molecules in Earth's atmosphere (basically, nitrogen and oxygen), they get excited, creating a chemical reaction that we see as light: greens and reds for oxygen, blue and purple-red for nitrogen. Aurorae move almost like waves, creating subtle curtains of light overhead. The phenomena can be seen on other planets and moons in the solar system.

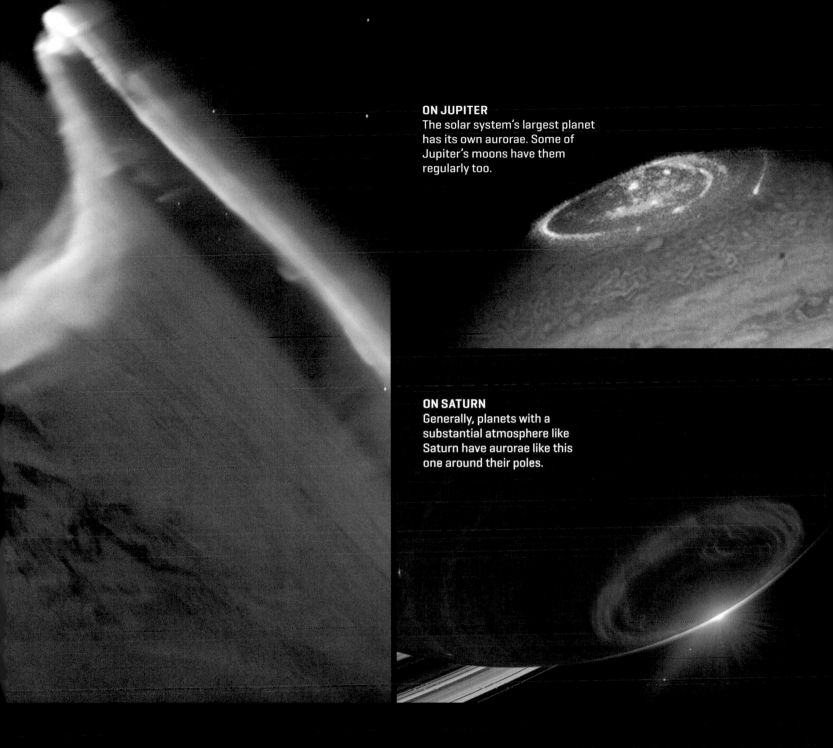

ON JUPITER

The solar system's largest planet has its own aurorae. Some of Jupiter's moons have them regularly too.

ON SATURN

Generally, planets with a substantial atmosphere like Saturn have aurorae like this one around their poles.

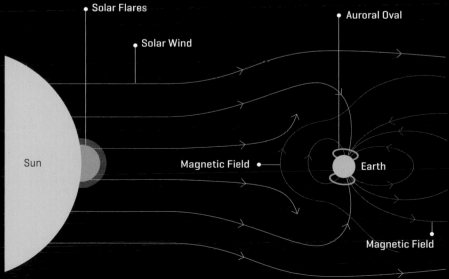

Solar Flares

Solar Wind

Auroral Oval

Sun

Magnetic Field

Earth

Magnetic Field

ENTRY CORRIDORS

The solar wind elongates the Earth's magnetosphere, sending streams of charged particles along its magnetic field lines. Those that make it through Earth's magnetosphere collide with chemicals in the atmosphere above the poles, creating beautiful aurorae. These phenomena happen some 100 kilometers (62.1 mi) above Earth's surface.

SOLAR ERUPTIONS

On occasion, especially when it is very active, the sun produces immense explosions accompanied by spectacular ejections of material. These solar eruptions, or flares, produce bursts of radiation across the electromagnetic spectrum that can last for minutes or hours.

Scientists classify solar flares according to their brightness in the x-ray wavelengths. There are three categories: X-class flares are major events that can trigger radio blackouts and radiation storms in Earth's upper atmosphere. M-class flares are medium size and can cause brief radio blackouts around the polar regions; minor radiation storms sometimes follow an M-class flare. A-, B-, and C-class flares are small, with few noticeable effects.

A Tight Window

Solar eruptions are observed through an H-alpha filter, which allows only radiation in a narrow range of wavelengths to pass through, seen in the red part of the electromagnetic spectrum. Images are almost always spectacular because they show the eruptions against the dark backdrop of space.

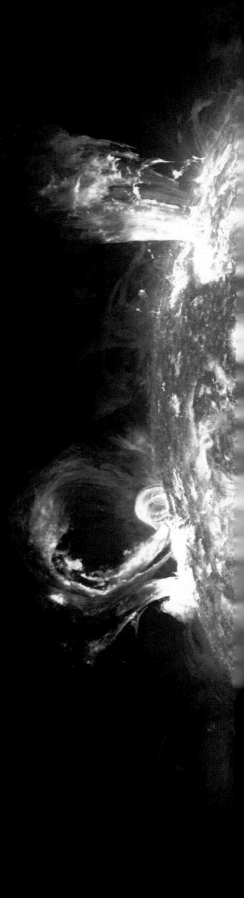

Types of Solar Flares

Each class of solar flare [A, B, C, M, and X] has a scale of 1 to 9, with 9 being the strongest flare of that class. Each class is 10 times more powerful than the one that comes before it, so an X2 solar flare is twice as powerful as an X1 and four times as powerful as an M5. The most powerful classes, M and X, are infrequent and linked to effects felt on Earth.

M2-type
Solar Flare

X2-type
Solar Flare

CLASSIFICATION	X-RAY FLOW PEAK
A	Less than 10^{-7} W/m^2
B	Between 10^{-7} and 10^{-6} W/m^2
C	Between 10^{-6} and 10^{-5} W/m^2
M	Between 10^{-5} and 10^{-4} W/m^2
X	More than 10^{-4} W/m^2

SEQUENCE OF SOLAR ERUPTIONS
This image combines photographs of solar eruptions captured by the Solar Dynamics Observatory over the course of a single year (2013–2014).

SOLAR FLARES AT DIFFERENT WAVELENGTHS

This sequence shows solar flares observed at different wavelengths, indicated in angstroms (Å). From left to right: the photosphere, chromosphere, unexcited corona, and active regions of the corona.

1,600 Å   304 Å   171 Å   335 Å   94 Å   131 Å

A SPECIAL LOOK AT ERUPTIONS
Filters let us capture images like this one of the sun's surface, composed of 3,336 photos taken by the Solar Dynamics Observatory (SDO). They make it easy to pick out flares from their surroundings, allowing the study of solar eruptions.

THE DANGERS OF SOLAR STORMS

Large solar storms affect Earth's magnetosphere, creating geomagnetic events that can impact satellites and Earth-based electrical grids.

Solar storms happen around coronal mass ejections and periods of high solar wind. They are sometimes intense enough to alter Earth's magnetosphere and cause so-called geomagnetic storms, which increase electric currents in the Earth's upper atmosphere and can result in an aurora. Solar storms can also disrupt global positioning systems, radar, high-frequency radio communications between aircraft and air traffic control, electrical grids, and communication technology that relies on satellites, such as cell phones. They can also affect the International Space Station.

Protective Measures

To prevent disruptive power outages on Earth, scientists monitor the sun's activities, especially when it comes to solar storms. When one is detected, satellites are reoriented or shut down and electrical grids are temporarily disconnected.

THREATS TO THE SATELLITE NETWORK

Powerful geomagnetic storms can affect satellites' electronics, putting them out of service either temporarily or permanently.

RISK TO ASTRONAUTS

Astronauts outside of Earth's magnetosphere on the International Space Station are less protected against solar storms, especially when they do spacewalks. Their bodies accumulate radiation during their time in orbit, which can lead to higher risks of developing cancer. As a precaution, they stay in the most protected zone of their spacecraft during solar storms.

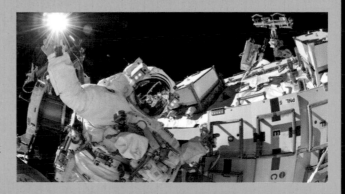

THE SOLAR CONSTANT

Earth's climate depends largely on how much energy our planet gets from the sun. Calculating this solar constant helps us measure its influence.

The amount of solar energy the Earth receives is called its solar constant, measured outside our atmosphere by calculating the light output over an area of one square meter oriented directly perpendicular to the sun. This energy per unit of time and area varies with the position of the Earth in relation to the sun, as Earth follows an elliptical orbit and its relative position changes over the course of a year. These variations do not determine the seasons: Earth is actually closer to the sun when the Northern Hemisphere experiences winter. We have seasons because our planet's axis of rotation is tilted with respect to our orbital plane (the plane of Earth's orbit around the sun).

The Primitive Sun

When our star was young, its luminosity reached only 70 percent of its current levels, with a solar constant that was correspondingly lower. Mysteriously, the temperature on Earth's surface was not chilly; water still existed as a liquid. This phenomenon is known as the "young sun paradox."

Current Solar Constant
Intensity at the Entry
Point of the Atmosphere
1,366 W/m²

Atmosphere
and exosphere:
about 8,000 km

Earth's diameter:
12,800 km

DETERMINING THE SOLAR CONSTANT

This value, measured outside the atmosphere, is determined by measuring the flow of solar radiation over a known area directly perpendicular to the direction of the sun.

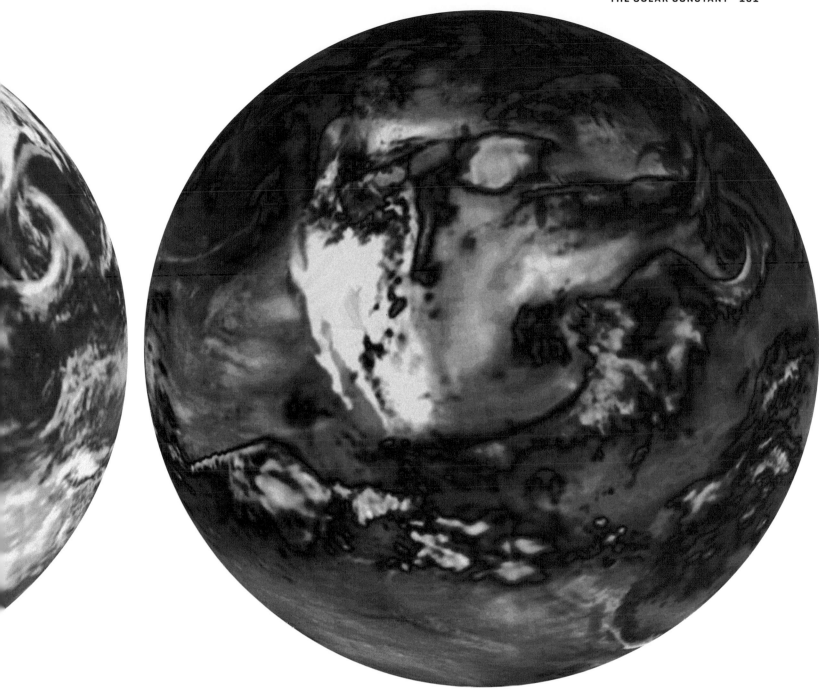

Solar Energy

When the sun's solar energy passes through our atmosphere, some of it is reflected back into space by clouds or Earth's surface. The rest is absorbed, trapped by greenhouse gases, heating up the planet's surface and emitting infrared radiation. This process is called the "greenhouse effect."

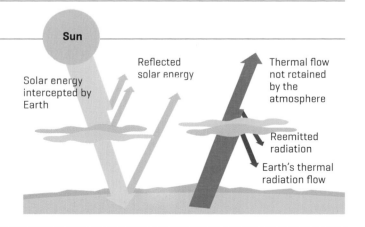

Sun

Solar energy intercepted by Earth

Reflected solar energy

Thermal flow not retained by the atmosphere

Reemitted radiation

Earth's thermal radiation flow

ENERGY EMITTED BY EARTH

These two images of Earth show how much solar light it reflects (at left) and how much heat escapes into space (at right), with the maximum amount of heat shown in yellow.

LIFE-GIVING HEAT
The sun is our main source of energy. Earth wouldn't be habitable without our star, which shaped our climate, allowed water to exist as a liquid, and created the perfect conditions for the development of complex living creatures.

MISSIONS TO THE SUN

Since 1960, space missions have investigated the sun
and its solar wind from both Earth's orbit and the sun's.

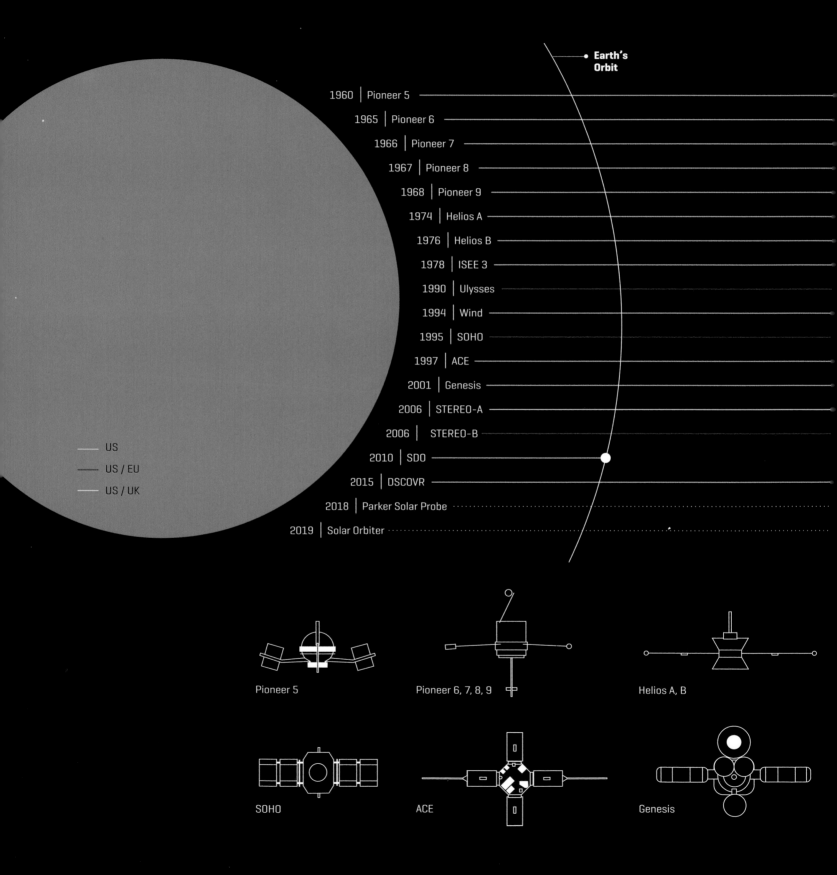

Earth's
Orbit

1960	Pioneer 5
1965	Pioneer 6
1966	Pioneer 7
1967	Pioneer 8
1968	Pioneer 9
1974	Helios A
1976	Helios B
1978	ISEE 3
1990	Ulysses
1994	Wind
1995	SOHO
1997	ACE
2001	Genesis
2006	STEREO-A
2006	STEREO-B
2010	SDO
2015	DSCOVR
2018	Parker Solar Probe
2019	Solar Orbiter

——— US
——— US / EU
——— US / UK

Pioneer 5

Pioneer 6, 7, 8, 9

Helios A, B

SOHO

ACE

Genesis

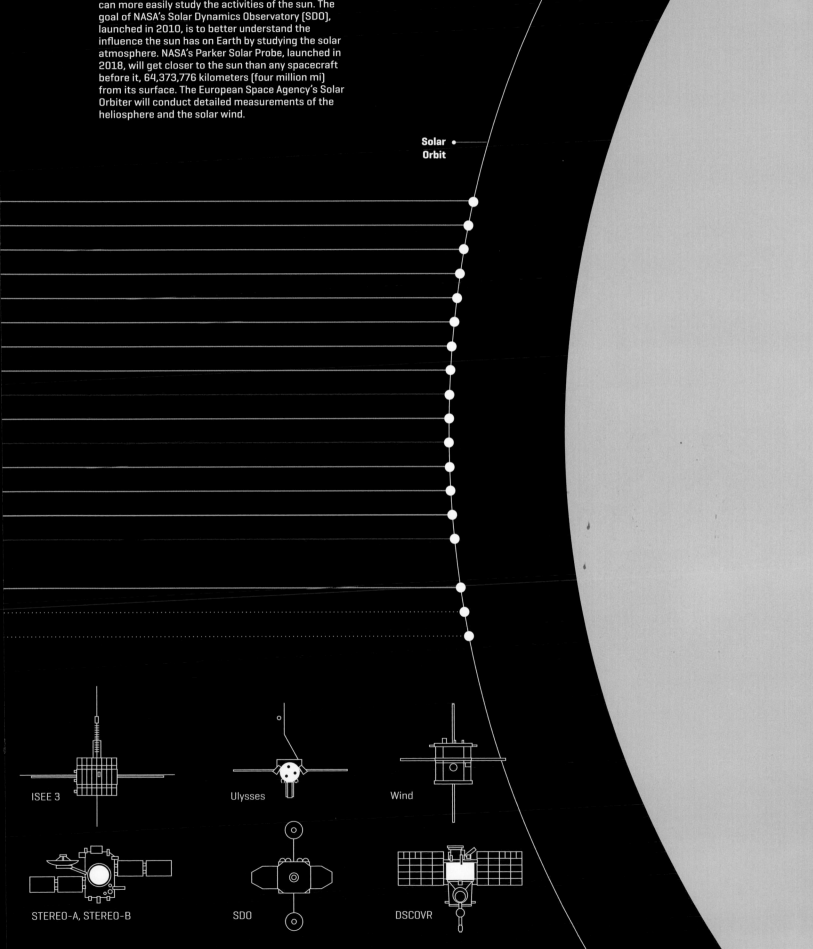

SOLAR PROBES
Free of the barrier of Earth's atmosphere, spatial observatories such as the Wind and Genesis probes can more easily study the activities of the sun. The goal of NASA's Solar Dynamics Observatory (SDO), launched in 2010, is to better understand the influence the sun has on Earth by studying the solar atmosphere. NASA's Parker Solar Probe, launched in 2018, will get closer to the sun than any spacecraft before it, 64,373,776 kilometers (four million mi) from its surface. The European Space Agency's Solar Orbiter will conduct detailed measurements of the heliosphere and the solar wind.

Solar Orbit

ISEE 3

Ulysses

Wind

STEREO-A, STEREO-B

SDO

DSCOVR

THE SOLAR

A HOME

SYSTEM:
FOR LIFE

THE FORMATION OF THE SOLAR SYSTEM

Our planetary system was created from material orbiting a young sun. This material began forming clumps that grew larger and larger, creating the planets as we know them.

The most accepted theory of how the planets formed is that cosmic dust orbiting the protosun began to form fragments called planetesimals. These planetesimals attracted other nearby particles and rapidly increased in size. Only a few grew to become oligarchs, or young precursors to today's planets. The oligarchs nearest the sun collided with each other and formed the inner or rocky planets, which are composed mostly of rocks and metals. Planetesimals farther away from the sun grew even larger, drawing in a larger number of planetesimals. These outer planets are composed mostly of gases, especially hydrogen and helium, with a relatively small rocky core. This process of planet formation took tens of millions of years.

A NEW PLANET

Pluto was considered the ninth planet until 2006, when it was downgraded to a dwarf planet. But the orbits of several distant objects suggest there is another planet in the solar system, although it has not yet been detected, at a distance of between 200 and 1,200 astronomical units (AU) from the sun. Its mass is estimated at 10 times that of Earth's.

FROM A CLOUD TO A PLANETARY SYSTEM

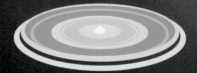

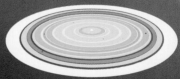

A Cloud of Dust and Gas
Some 4.6 billion years ago, a very energetic event—most likely a nearby supernova—forced a large cloud of gas and dust to contract.

Creating a Star
This contraction created a flat, rotating disc of gas and dust. Gravity pushed material into its center, creating a protosun, which would end up accumulating 99.9 percent of the disc's total mass.

Sun King
When the core temperature of the protosun reached a critical point, hydrogen atoms began to fuse, releasing large amounts of energy and creating helium atoms, forming the sun as we know it.

The Planets Form
Most of the material not absorbed by the protosun clumped into larger and larger fragments called planetesimals: the seeds of the planets. The rest became moons, asteroids, and comets.

HL Tauri: A Planetary System Forming

HL Tauri, a star 450 light-years from Earth in the Tauri molecular cloud and barely 100,000 years old, is currently creating a new stellar system. This amazing image from the ALMA radio telescope shows the star in the center of a disc of dust and gas, whose dark furrows could indicate the presence of young planets.

The Early Solar System
The solar wind pushed the lightest substances away from the protosun. While the planetesimals farther from the protosun gained mass and attracted large amounts of gas, thanks to low temperatures, those closer in came together with heavy materials to form small, rocky orbs.

A MATURE SYSTEM
In this illustration of the formation of a planetary system, the protoplanetary disc has given most of its mass away to the star and the planets.

THE SIZE OF THE SOLAR SYSTEM

Most of the solar system's mass is split between the sun (more than 99 percent) and the eight planets and their 175 moons.

The Solar System's Masses

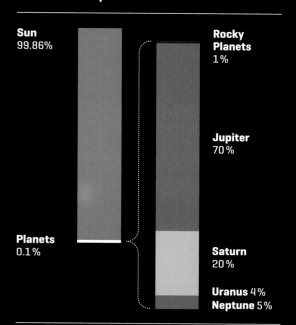

Sun
99.86%

Planets
0.1%

Rocky Planets
1%

Jupiter
70%

Saturn
20%

Uranus 4%
Neptune 5%

Saturn

Jupiter

Mars

Earth

Phobos
Deimos

Venus

Moon

Mercury

	MERCURY	VENUS
Distance to the sun	57.9 million km [36 million mi]	108.2 million km [67.2 million mi]
Rotation (Earth time)	587 days	243 days
Orbital period (Earth time)	88 days	225 days
Diameter	4.879 km [3.031 mi]	12.104 km [7.521 mi]
Mass [× Earth]	3.302×10^{23} kg [0.06]	4.869×10^{24} kg [0.8]
Atmosphere	O_2. Na. H_2	CO_2. N_2
Moons (main)	0	0

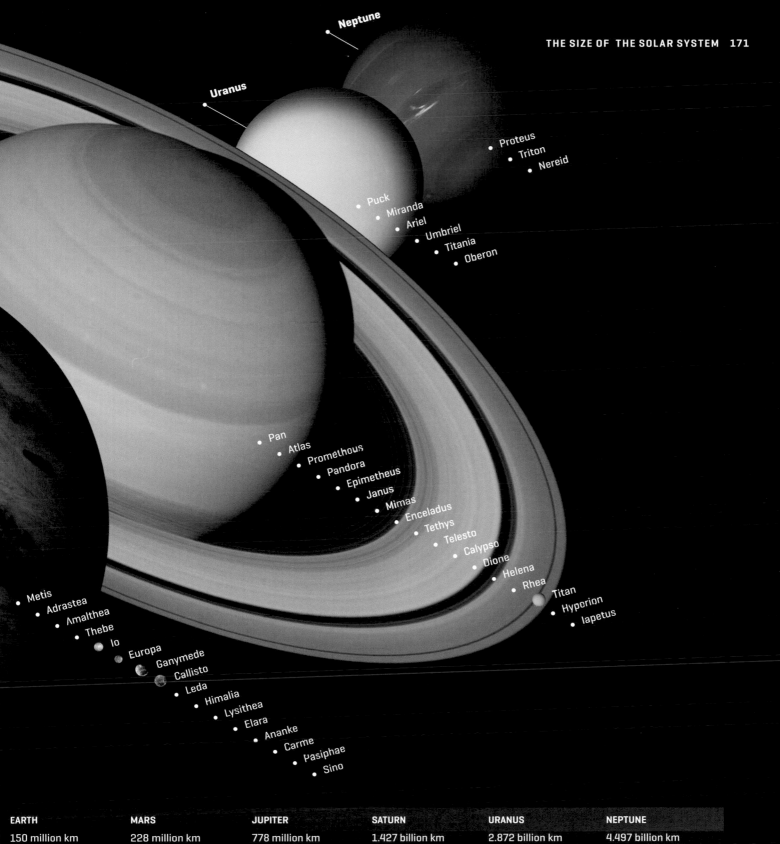

Neptune

Uranus

Proteus
Triton
Nereid

Puck
Miranda
Ariel
Umbriel
Titania
Oberon

Pan
Atlas
Prometheus
Pandora
Epimetheus
Janus
Mimas
Enceladus
Tethys
Telesto
Calypso
Dione
Helena
Rhea
Titan
Hyperion
Iapetus

Metis
Adrastea
Amalthea
Thebe
Io
Europa
Ganymede
Callisto
Leda
Himalia
Lysithea
Elara
Ananke
Carme
Pasiphae
Sino

EARTH	MARS	JUPITER	SATURN	URANUS	NEPTUNE
150 million km (93 million mi)	228 million km (142 million mi)	778 million km (484 million mi)	1.427 billion km (887 million mi)	2.872 billion km (1.784 billion mi)	4.497 billion km (2.794 billion mi)
1 day (23 h 56 min)	24 h 37 min	9.8 hours	10.55 hours	17 hours	16 hours
1 year (365.25 days)	687 days	11.86 years	29.45 years	84.3 years	164.78 years
12,756 km (7,926 mi)	6,787 km (4,222 mi)	142,796 km (88,729 mi)	120,660 km (74,600 mi)	51,118 km (32,600 mi)	49,528 km (30,775 mi)
5.972×10^{24} kg [1]	6.42×10^{23} kg [0.1]	1.898×10^{27} kg [318]	5.688×10^{26} kg [95]	8.681×10^{25} kg [14.5]	1.024×10^{26} kg [17]
O_2, N_2	CO_2, N_2, Ar	H_2, He	H_2, He	H_2, He, CH_4	H_2, He, CH_4
1 [Moon]	2	79 [Io, Europa, Ganymede, Callisto]	62 [Titan]	27	14

DISTANCES IN THE SOLAR SYSTEM

Although the solar system is approximately 15 trillion kilometers (9.32 trillion mi) across, its mass is concentrated in an area of 4.5 billion kilometers (2.8 billion mi), including the sun and the planets that orbit it.

Mars
1.52 AU

Earth
1 AU

Inner solar system
Planets inside the asteroid belt are composed of rocks and metals, the only materials that can resist such high temperatures.

Neptune
30 AU

Uranus
19.2 AU

FROM THE SUN TO PLUTO
The distances between planets in our solar system are so vast that it isn't practical to measure them in kilometers or miles. Instead, distances are measured in astronomical units (AU). One AU is equal to the approximate distance between Earth and the sun: 150 million kilometers, or 93 million miles. The distance from the sun to Pluto is 39.5 AU.

Saturn
9.5 AU

THE SOLAR SYSTEM BY DISTANCE

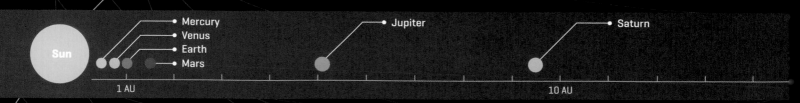

Sun · Mercury · Venus · Earth · Mars · Jupiter · Saturn

1 AU   10 AU

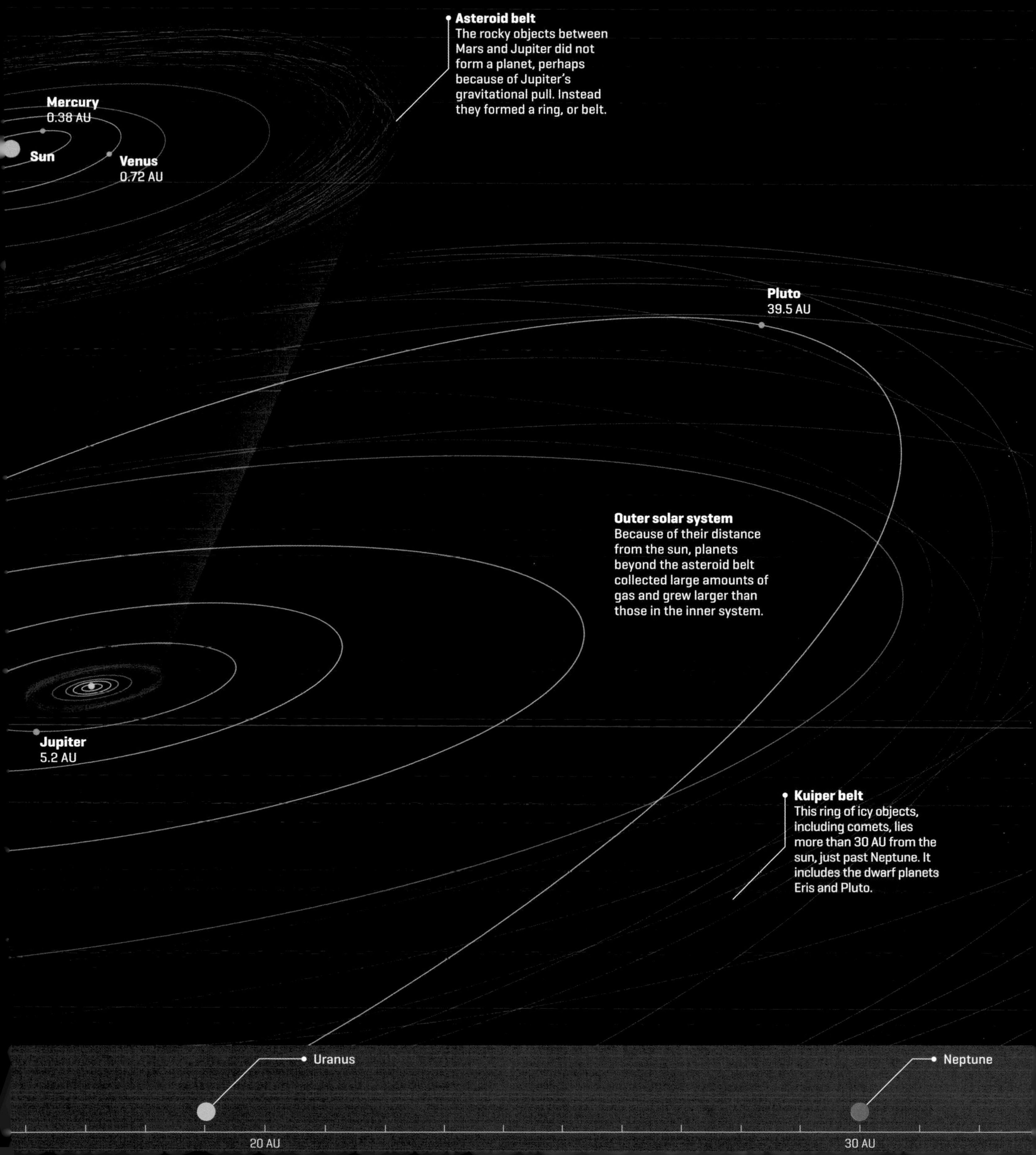

Asteroid belt
The rocky objects between Mars and Jupiter did not form a planet, perhaps because of Jupiter's gravitational pull. Instead they formed a ring, or belt.

Mercury
0.38 AU

Sun

Venus
0.72 AU

Pluto
39.5 AU

Outer solar system
Because of their distance from the sun, planets beyond the asteroid belt collected large amounts of gas and grew larger than those in the inner system.

Jupiter
5.2 AU

Kuiper belt
This ring of icy objects, including comets, lies more than 30 AU from the sun, just past Neptune. It includes the dwarf planets Eris and Pluto.

Uranus

Neptune

20 AU

30 AU

ORBITS: LASTING BONDS

An orbit is the path an object follows around a star, planet, or moon, elliptical in shape and influenced by gravitational attraction. Its regularity and continuity obey simple but powerful laws.

Supported by observational data gathered by astronomer Tycho Brahe (1546–1601), his German colleague Johannes Kepler (1571–1630) wrote three famous laws that elegantly described the orbits of the planets around our sun. The first law says that planets move along elliptical orbits with the sun at one focus, discarding the perfect circular orbits inherited from the ancient Greeks.

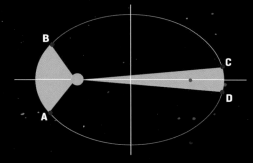

The second law states that if you draw a line from the sun to a planet and wait a fixed amount of time, the line will sweep equal areas in equal times. In other words, the area is the same no matter where in the orbit you are. The planet takes the same amount of time to travel between points A and B (close to the sun) as C and D (farther away) because the closer it is, the faster it moves. Think of a stone thrown in the air: it loses speed as it rises to the top of its trajectory, then speeds up again as it falls.

The third law affirms that the relation between the orbital period (T), or the time it takes to complete one orbit, and the semi-major axis of the ellipse (r) is constant. Scientists already knew the Earth's orbital period and its distance to the sun. They were then able to calculate the orbital periods of every planet using Kepler's third law to figure out distances in the solar system.

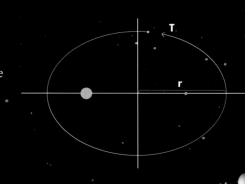

THE DANCE OF GRAVITY
This illustration shows one of gravity's most impressive feats: its ability to hold onto Saturn's orbital system, its rings of tiny rocks and ice crystals, and Thetis, one of 60 moons.

SIGNS OF LIFE ON ENCELADUS?
In 2005, the Huygens probe discovered that
Saturn's icy moon is geologically active. The
immense geysers at the south pole eject ice
with traces of hydrogen, as illustrated here.
The presence of this element suggests that
they are hydrothermal vents and, like the
ones on Earth, could contain microbial life.

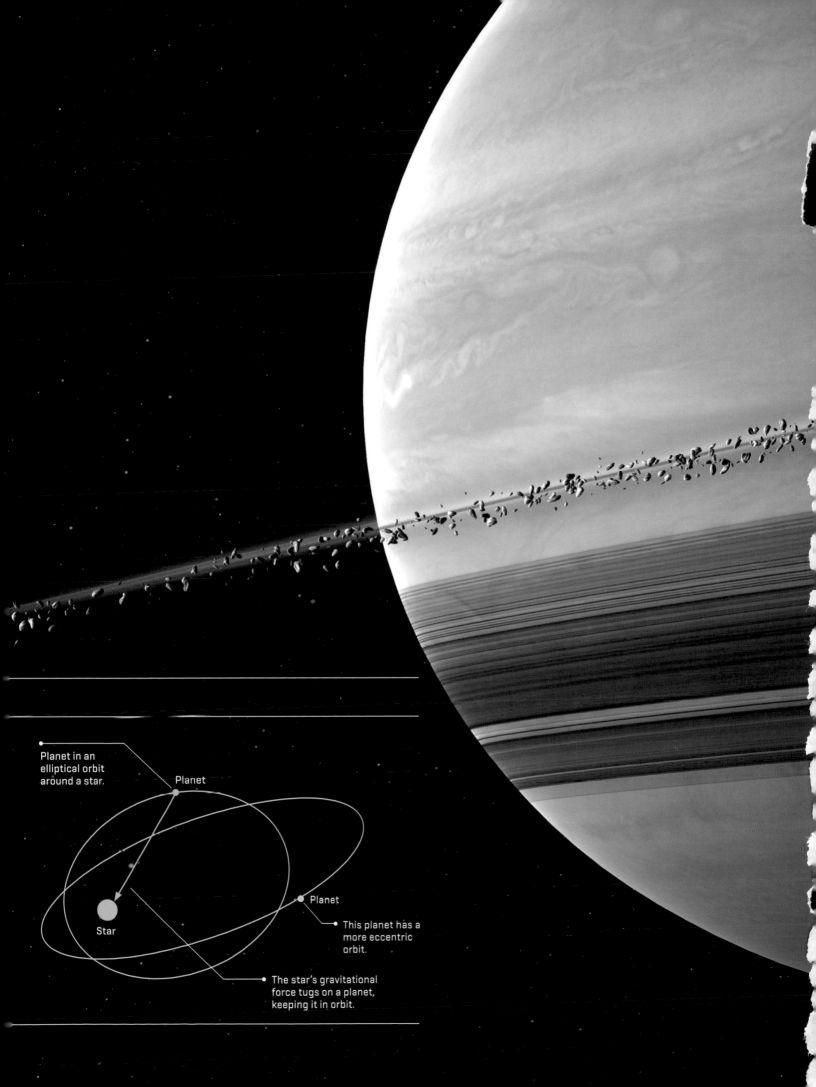

Planet in an
elliptical orbit
around a star.

Planet

Planet

This planet has a
more eccentric
orbit.

Star

The star's gravitational
force tugs on a planet,
keeping it in orbit.

ORBITAL INCLINATION

The planets are often shown in beautiful alignment because they have the same orbital inclination: in other words, the same tilt as they circle the sun. To understand this phenomenon, we have to go back to the origins of the solar system.

The fact that the planets share a common past, formed in the cradle of the protoplanetary disc, has a series of notable consequences. Perhaps the most notable is the similarity between the tilt of the planets' respective orbits. If we were to draw a plane connecting the center of the sun and Earth, with the Earth in two different positions along its orbit, we would have Earth's orbital plane, or ecliptic. If we take the weighted average of the mass and angular moments of each planet, we would have the invariable plane. The incline of the different orbits with respect to the invariable plane is tiny. Various phenomena occur because of this alignment, either between themselves or with respect to the sun. The movements of planets in relation to the sun, and each other, are the basis for the zodiac.

Orbital Plane and Earth's Equator

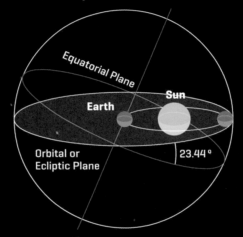

Mars

Saturn

DEFINITIONS
The orbital or ecliptic plane is the imaginary line that traces the path of the sun across the sky. An orbital inclination is the angle of a planet or other body to the ecliptic plane. The invariable plane is the average plane of the orbits of the planets. An equator is the imaginary line around the middle of the Earth or other body, such as the sun or other planet, that is equally distant from its north and south poles. An equatorial plane is the plane passing through the equator of the Earth or other body.

ORBITAL DEVIATIONS FROM THE INVARIABLE PLANE

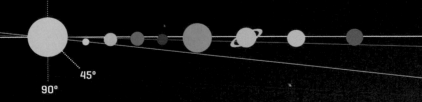

90°

45°

Invariable Plane: 0 degrees
Ecliptic Plane: 1.6 degrees

Mercury's Plane: 6.3 degrees

The planet's degree of orbital deviation with respect to the invariable plane reveals that only Mercury has a steep incline: a bit more than 6 degrees.

Mercury	6.3°	Uranus	1.0°
Venus	2.2°	Saturn	0.9°
Mars	1.7°	Neptune	0.7°
Earth	1.6°	Jupiter	0.3°

Constellations of the Zodiac

The zodiac is an imaginary band in the sky that extends about 9 degrees on either side of the sun's path, or its ecliptic plane. The movements of the sun, the planets, and all or part of 24 constellations can be found within it. During the year, the sun crosses 13 of these constellations: the 12 traditional signs of the zodiac, plus Ophiuchus (the Serpent-bearer).

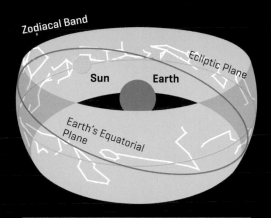

Zodiacal Band

Ecliptic Plane

Sun   Earth

Earth's Equatorial Plane

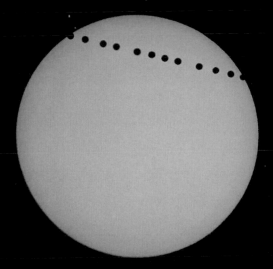

THE TRANSIT OF VENUS

A transit happens when a relatively small object passes directly across a larger body, eclipsing only a very small area. From Earth, the most notable transits occur when Mercury and Venus pass in front of the sun. The transit of Venus was especially useful to 17th-century astronomers as a means of accurately measuring the distance between the Earth and the sun.

Mercury

THE PLANTARY PROCESSION
This image, taken in 2006 from the Spanish island of Tenerife, shows the alignment of three of the five planets we can see with the naked eye on an almost perfect diagonal: Mars, Saturn, and Mercury.

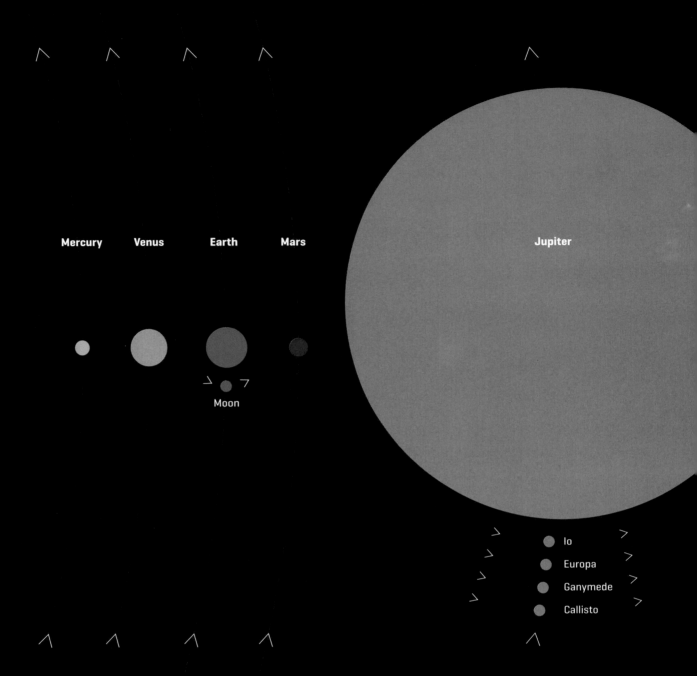

Mercury   Venus   Earth   Mars   Jupiter

Moon

Io
Europa
Ganymede
Callisto

OTHER CONSTANTS IN THE SOLAR SYSTEM

The planets all orbit in the same plane, and they also orbit in the same direction. The explanation for their constancy goes back to their beginnings as part of the protoplanetary disc.

If we observe the planets from Earth's North Pole, we can see that they all rotate counterclockwise— the same direction the protoplanetary disc once rotated. This is also true of larger moons, which presumably broke away from planets due to a violent impact. This common direction of rotation is called prograde motion.

Saturn

Uranus

Neptune

Titania

Oberon

Triton

Rhea

Titan

Iapetus

PLANETARY ROTATION

Most of the planets in our solar system rotate in the same direction: all except for Uranus and Venus. They have a sharply inclined axis of rotation, probably as a result of being hit by other large bodies during their formation.

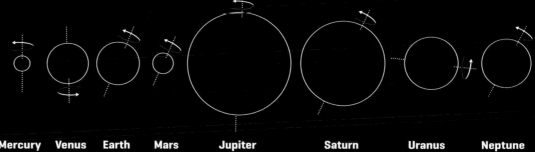

Mercury	Venus	Earth	Mars	Jupiter	Saturn	Uranus	Neptune
~0°	177.36°	23.44°	25.19°	3.13°	26.73°	97.77°	28.32°

WHAT IS A PLANET?

Determining whether or not a body is a planet comes down to its mass and relationship to the objects around it, as well as its closest star.

The word *planet* comes from the Greek *planetes*, which means "wanderer." The ancients thought the five planets visible to the naked eye—Mercury, Venus, Mars, Jupiter, and Saturn—all orbited Earth, the sun and moon along with them. But what does *planet* mean in astronomical terms? The International Astronomical Union defines it as a celestial body that meets three key criteria.

Orbiting a Star

The solar system contains celestial bodies that are similar in size to the planets but that are not considered planets because they do not orbit the sun. Good examples are two of Jupiter's moons, Ganymede and Callisto, which are both larger than Mercury, and Saturn's largest moon, Titan.

Practically Spherical

A planet must be spherical, or nearly so. That happens when a celestial body has enough mass that gravity is in balance with the force of its internal pressure, creating hydrostatic equilibrium. There are a multitude of smaller, nonspherical bodies orbiting the sun, such as most of the objects in the asteroid belt between the orbits of Mars and Jupiter.

Orbital Dominance

To count as a planet, a body must have orbital dominance: in other words, enough gravitational force to clear its galactic neighborhood of debris. Bodies that satisfy all criteria except this one, such as Pluto, are called dwarf planets.

THE LOCAL GIANT
Jupiter is a true colossus: its mass is more than twice the rest of the planets' mass put together. As such it has a lot of gravitational influence, which would have impacted how the solar system formed.

PLUTO IN DETAIL
This color image of Pluto highlights differences in its chemical makeup at the surface; for example, areas rich in methane and nitrogen are shown in dark brown.

PLUTO AND OTHER DWARF PLANETS

The dwarf planet category was created in 2006 by the International Astronomical Union to help categorize bodies that ticked most of the planetary boxes, but had not cleared their immediate vicinity of debris. Pluto was considered a planet from its discovery in 1930 until 2006, despite its eccentric inclined orbit and the revelation in 1978 that it had barely 5 percent of Mercury's mass. Another body that falls into dwarf planet category is Ceres, the largest of the known asteroids, and other bodies in the Kuiper belt or farther out, including Eris, Makemake, and Haumea.

Planets Are Not Stars

The International Astronomical Union specifies that a planet cannot emit its own light, although all bodies emit some—a phenomenon that increases as temperatures rise. What truly differentiates a planet and a star is whether it has gathered enough mass to produce continuous hydrogen fusion. Jupiter and the sun are both made mostly of hydrogen, but Jupiter's mass is some 1,000 times less than the sun's. If the sun's mass were added to Jupiter, its increased density would also increase its gravitational pressure and decrease its size. Once it reached 13 times its current mass, it would be hot enough to begin physicochemical processes that would transform it into a brown dwarf. Once it formally became a star, hydrogen fusion would only continue if the mass were 80 times its current value. Below are some stars between the average size and Jupiter according to their mass and including their temperatures.

Small to medium size star 1 solar mass	5530 °C (9986°F)	Sun
Star of minimal mass (red dwarf) less than 0.5 solar mass	3530 °C (6386°F)	Gliese 229A
Young brown dwarf 13–80 Jupiter masses	2225 °C (4037°F)	Teide 1
Ancient brown dwarf 13–80 Jupiter masses	2025 °C (3677°F)	Gliese 229B
Isolated bodies of planetary mass 10 Jupiter masses	1225 °C (2237°F)	Sigma Ori 70
Planet 1 Jupiter mass	-90 °C (-130°F)	Jupiter

PLANETARY COMPOSITIONS

The inner planets—Mercury, Venus, Earth, and Mars—are made mostly of rocks and metals. That's why they are called rocky or terrestrial planets. The outer planets—Jupiter, Saturn, Uranus, and Neptune—are mostly gas and ice.

The rocky planets have iron-rich metallic cores, mantles, and hearts of rock. Gas giants are made mostly of hydrogen and methane, although they can have small, solid cores that include rocks and hydrogen compounds like water. Rocky planets have a well-defined, solid surface, while gas giants have surfaces that shift from gas to liquid to solid states as the atmosphere thickens. Their compositions also impact their gravity: despite having a much lower mass, rocky planets have a higher density, which makes their gravity similar to that of the gas giants. Their composition is directly related to where they formed in the protoplanetary disc: rocky planets are found in the inner solar system, while gas giants are farther from the sun.

RED TUNIC
Most of the Martian surface is made of basalt, but a thick layer of iron oxide dust is what gives it its characteristic red hue.

ROCKY PLANETS

Moon (to scale)

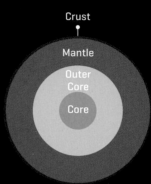

Earth

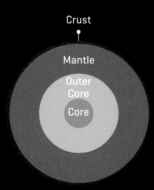

Venus

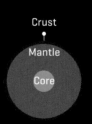

Mars

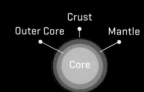

Mercury

- Molten Rock
- Rock
- Molten Iron
- Solid Iron
- Solid Iron (rich in iron sulfide)

GAS PLANETS

Earth (to scale)

Outer Atmosphere

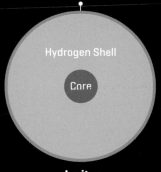

Hydrogen Shell

Core

Jupiter

Outer Atmosphere

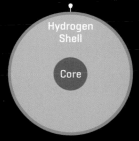

Hydrogen Shell

Core

Saturn

Shell

Core

Mantle

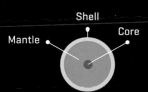

Uranus

Shell

Core

Mantle

Neptune

- ● HydrIce
- ● Rock/Ice
- ● Methane
- ● Metallic Hydrogen
- ● Atmosphere

Satellites and Rings

A defining characteristic of the gas giants is that they have large moons and ring systems. Saturn has the largest number of moons, followed by Jupiter, Uranus, and Neptune. Only the largest moons have a spherical shape. Hazy, huge Titan is covered with liquid hydrocarbons, while one of Enceladus's poles has cryovolcanic geysers that blast water into space. Both could be hiding water beneath their surface that could, potentially, hold some form of life.

A GARLAND OF PARTICLES
Saturn's rings stretch across 282,000 kilometers (175,000 miles), but they're thin—only a few stories tall—and full of moons and smaller bodies that carve pathways through them. This image of Saturn's rings, taken by the Cassini probe, shows the varying size of the particles within them. The biggest ones are 5 centimeters (about 2 in) or more, shown in pink, while the blue and green ones are smaller. The smaller the particles, the bluer they appear.

THE PLANETS ALIGN
Once the planets formed, interactions
between them gave rise to the solar
system's shape. The outer gas planets—
Jupiter, Saturn, Uranus, and Neptune—
moved farther away from the sun due
to their gravitational forces. Many of the
fragments of rock and ice that originally
surrounded them moved to become the
inner rocky planets—Mercury, Venus, Earth,
and Mars.

OBSERVING THE SOLAR SYSTEM

1610

Galileo Galilei used a primitive telescope to observe sunspots, Jupiter's moons, and, most importantly, the moon's irregular surface with its valleys and mountains, which he drew in his work *Sidereus nuncius* (*Sidereal*, or *Starry, Messenger*).

1665

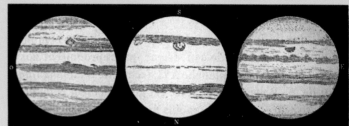

Giovanni Cassini discovered that Jupiter rotated around its own axis, determined that rotation's duration, and drew the Great Red Spot as he saw it through his telescope.

1977

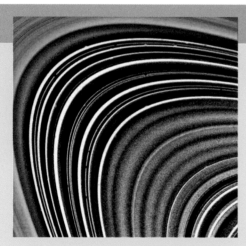

The Voyager 1 and **2** probes were launched to take advantage of the extremely rare alignment of Jupiter, Saturn, Uranus, and Neptune, exploring the gas planets for the first time. Both probes are now outside the solar system, although they are still transmitting valuable data about their environments.

1986

The core of **Halley's comet** was photographed by the Giotto probe, the only probe to accomplish the feat. Five other probes failed, although one of them, called ICE, entered into the comet's environment.

2010

The Solar Dynamics Observatory and its predecessor, the **SO HO** (1995), revolutionized our knowledge of solar dynamics and captured some of the sun's most spectacular ejections.

2012

The Mars rovers, some active since 2004, have supplied memorable images like a self-portrait of Opportunity on the Endeavour crater's eastern edge. The most recent rover, InSight, touched down on Mars in November 2018.

From the illustrations of what Galileo and Cassini saw through their telescopes to the sophisticated images obtained by modern space exploration probes, our observations of the solar system have come in giant leaps. These are some of space exploration's major milestones.

1845

The **first photograph of the sun** was taken by physicists Léon Foucault and Hippolyte Fizeau.

1959

The Soviet probe **Luna 3** photographed the **dark side of the moon** for the first time, shining some light on what the far side of the moon was like.

1976

The **Viking 2 probe** separated from the Viking spacecraft and landed on Utopia Planitia on Mars, where it captured the first photographs ever taken from the surface of Mars.

1989

The **Galileo probe** obtained new photos of Jupiter's four largest satellites, the so-called Galilean moons, such as Io, revealing its volcanoes in detail.

1990

The **Hubble Space Telescope** captured time-dependent phenomena such as the polar aurorae on Saturn and Jupiter. Hubble remains in operation today.

1997

The **Cassini-Huygens mission**, a collaboration between NASA, the European Space Agency, and the Italian Space Agency, got up close and personal with Saturn's rings and main moons. After taking over 453,000 images of the planet, the probe was destroyed in 2017.

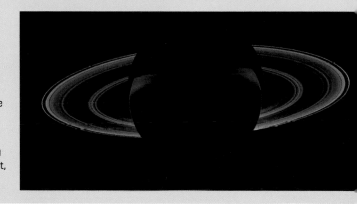

2015

The **Rosetta probe** pulled into orbit around a comet, 67P/Churyumov-Gerasimenko, and stayed there for a year. It took detailed photos as it approached.

2015

Launched in 2006, the **New Horizons probe** conducted a six-month study of Pluto and its moons on its way to the Kuiper belt and other trans-Neptunian objects.

2017

The **Juno probe** began collecting data on Jupiter that, in combination with other sources like the Earth-based Gemini telescope, is providing new insights into the gas giant and its atmosphere.

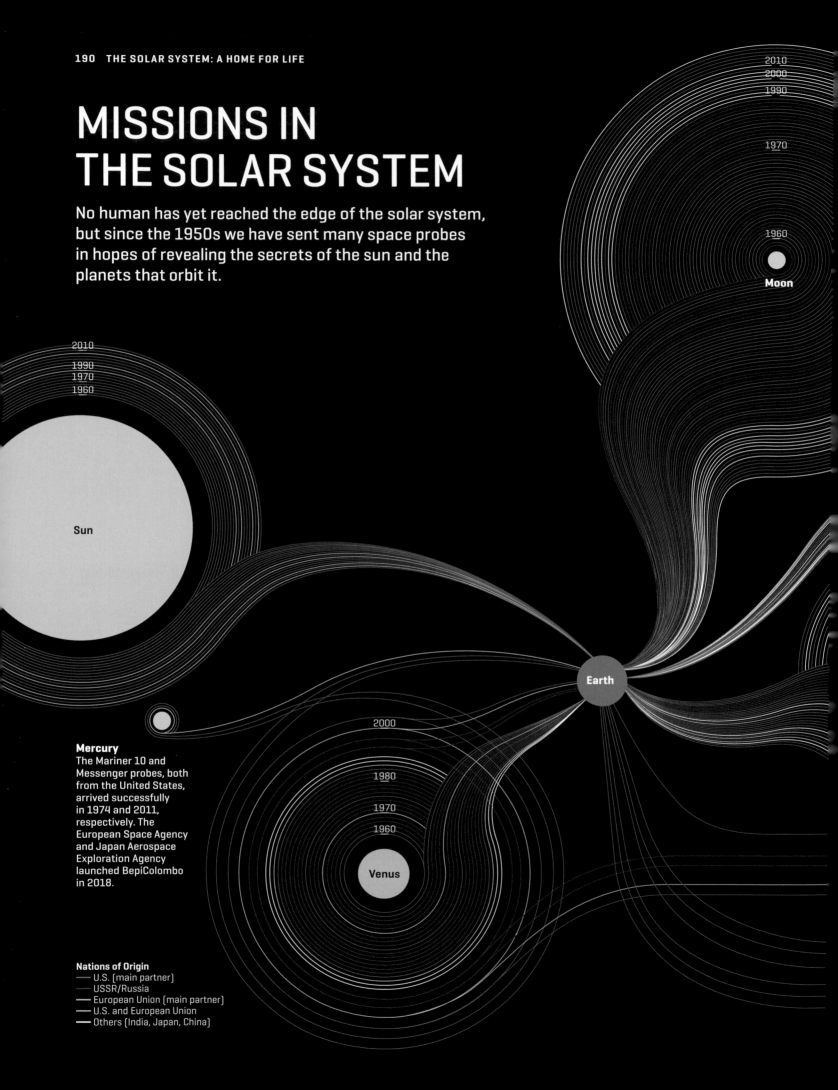

MISSIONS IN THE SOLAR SYSTEM

No human has yet reached the edge of the solar system, but since the 1950s we have sent many space probes in hopes of revealing the secrets of the sun and the planets that orbit it.

2010
2000
1990
1970
1960

Moon

2010
1990
1970
1960

Sun

Earth

2000

1980

1970

1960

Venus

Mercury
The Mariner 10 and Messenger probes, both from the United States, arrived successfully in 1974 and 2011, respectively. The European Space Agency and Japan Aerospace Exploration Agency launched BepiColombo in 2018.

2010

Nations of Origin
— U.S. (main partner)
— USSR/Russia
— European Union (main partner)
— U.S. and European Union
— Others (India, Japan, China)

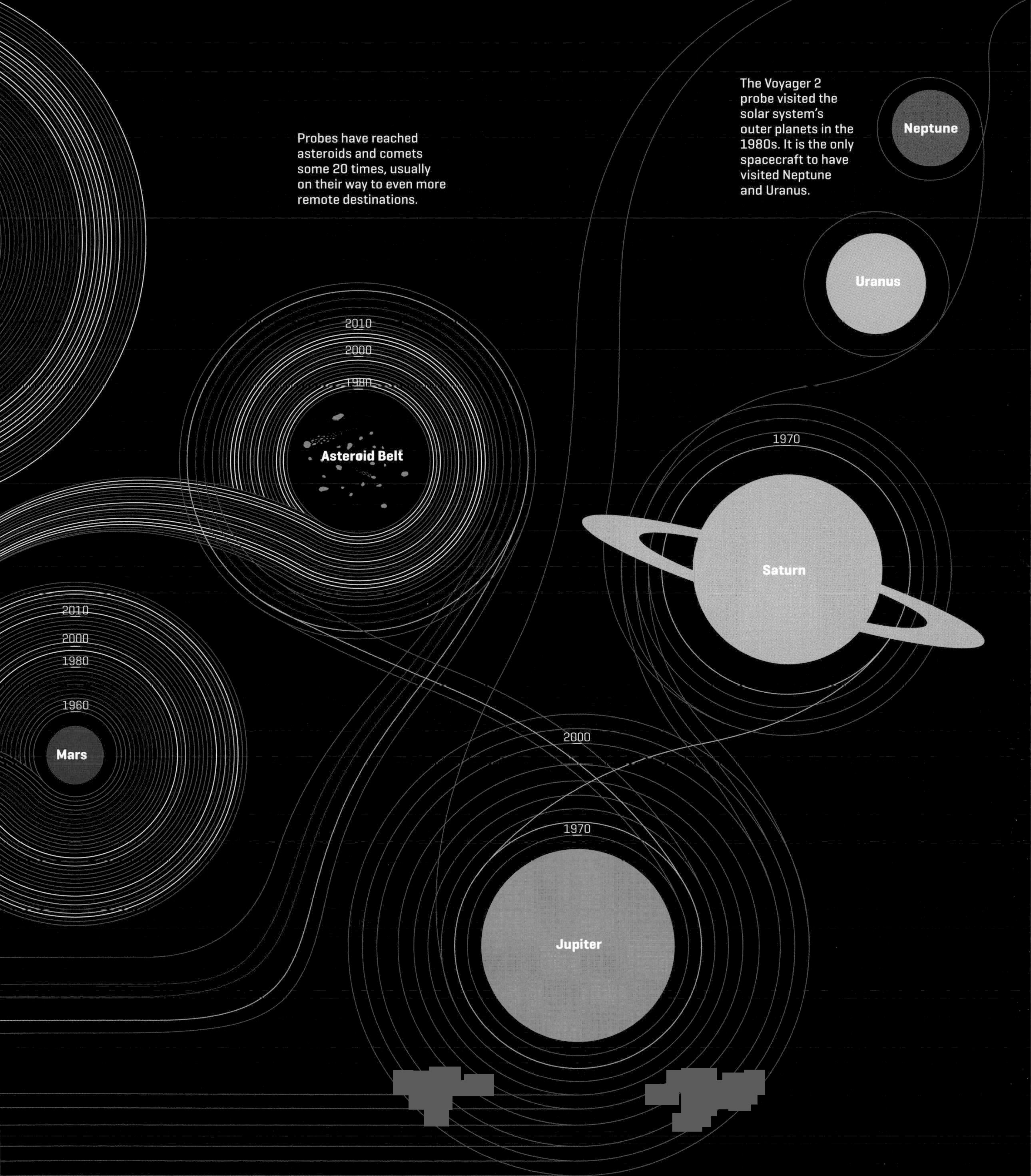

The Voyager 2 probe visited the solar system's outer planets in the 1980s. It is the only spacecraft to have visited Neptune and Uranus.

Probes have reached asteroids and comets some 20 times, usually on their way to even more remote destinations.

Neptune

Uranus

2010
2000
1980

Asteroid Belt

1970

Saturn

2010
2000
1980

1960

Mars

2000

1970

Jupiter

THE SCIENCE OF INTERPLANETARY FLIGHT

Moving through the solar system requires a lot of navigational precision, including the ability to discern when to fight against gravitational forces and when to take advantage of them.

The most general theory of the physical world is relativity, but at speeds much slower than the speed of light, like those achieved by current probes (including the fastest, the Helios-A probe), the laws of Newtonian gravity apply. The first hurdle for space missions is escaping Earth's gravity. To overcome it, spacecraft have to be launched at the so-called escape velocity, which for Earth is some 11.2 kilometers (7 mi) per second or over 40,000 kilometers (25,000 mi) per hour. Once they reach this speed, they have to deal with solar gravity. The escape velocity from the sun, initiated

from Earth's orbit, is 42.1 kilometers (26 mi) per second or a bit more than 150,000 kilometers (93,206 mi) per hour. As a probe reaches or moves away from its destination it has to keep adjusting its speed, depending on the gravity it's dealing with.

An Orbital Ally
The prograde motion of the planets' orbits means that probes can pass unharmed between them. Probes can use a single trajectory to visit different planets, as the Cassini probe did on its trip to Saturn.

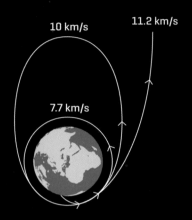

Escaping Earth
The speed necessary to orbit Earth is less than needed to escape it: 7.7 kilometers (4.8 mi) per second, or 25,920 kilometers (16,106 mi) per hour. Intermediate velocities cause more or less elliptical orbits.

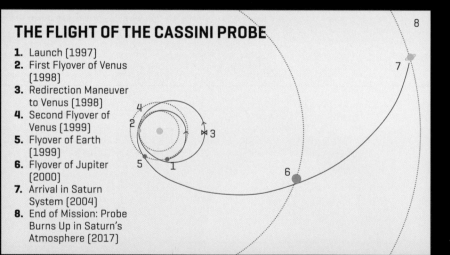

THE FLIGHT OF THE CASSINI PROBE

1. Launch (1997)
2. First Flyover of Venus (1998)
3. Redirection Maneuver to Venus (1998)
4. Second Flyover of Venus (1999)
5. Flyover of Earth (1999)
6. Flyover of Jupiter (2000)
7. Arrival in Saturn System (2004)
8. End of Mission: Probe Burns Up in Saturn's Atmosphere (2017)

VOYAGES IN SPACE AND TIME

Speed of Light (299,792 km or 186,282 mi/s)

Helios-A probe (252,793 km/h)

Moon   Venus   Mars   Jupiter   Saturn   Uranus   Pluto   Eris   Heliopause
Neptune

1 second   1 minute   1 hour   1 day

DISTANT TRAVELERS
NASA launched Voyager 1 and Voyager 2 in 1977 to take advantage of the alignment of Jupiter, Saturn, Uranus, and Neptune. Voyager 2 used prograde motion to visit the four gas giants. Both left the solar system, taking samples of humanity's art and science with them.

Speeding Up and Slowing Down

+ Speed −

Ship

When an object like a probe enters the elliptical orbit of a much larger body, like a planet, there is an exchange of kinetic energy that allows the smaller object to speed up or slow down without using fuel. To accelerate, the angle of incidence with respect to the planet's trajectory has to be larger than that of escape (shown in red); to brake, it should be less (in blue).

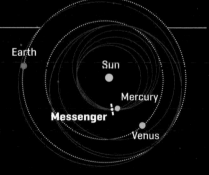

Earth
Sun
Mercury
Messenger
Venus

Slow Motion
To enter a stable orbit around Mercury, the Messenger probe reduced its velocity by doing successive orbits around Earth, Venus, and finally around Mercury itself.

Alpha Centauri · Betelgeuse · Center of the Galaxy

1 month · 1 year · 1 decade · 1 century · 1 millennium · 10,000 years · 100,000 years · 1 million years

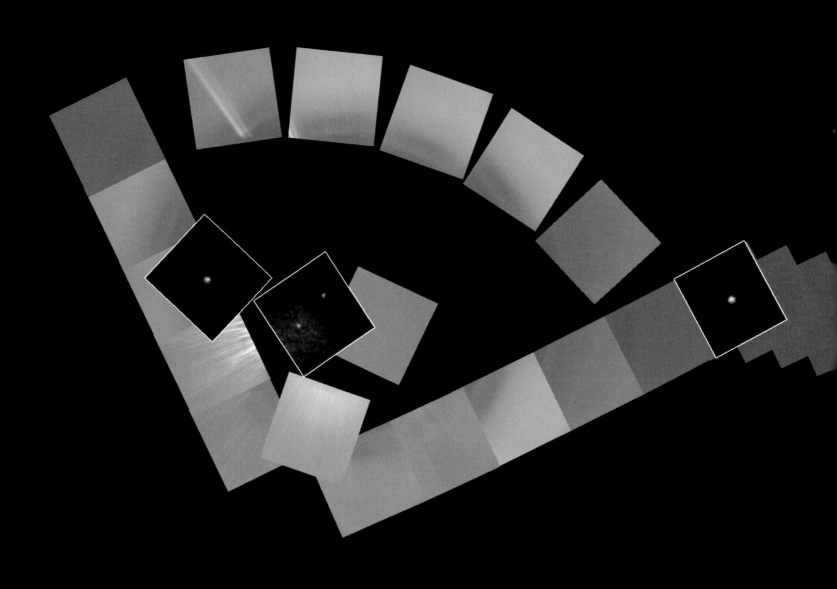

THE SOLAR FAMILY
On February 14, 1990, Voyager 1 took
one last look back at the solar system.
The result was this image, known as the
"Family Portrait." Made of 60 individual
photographs, its serpentine form reflects
its successive adjustments so it could
frame the planets. In the portrait, from left
to right: Jupiter, Earth and Venus (which,
due to their proximity, were captured in
one photo), Saturn, Uranus, and Neptune.

GAS PLANETS, ROCKY PLANETS

All planets originated from the protoplanetary disc that once surrounded the sun. Their closeness to the sun determined which bodies would go on to form a planet and shaped their type: gas giant or rocky planet.

If we could travel into the past, to the moment our star was being formed, we would see it surrounded by a turbulent disc of dust and gas. Our protostar accumulates mass through accretion, or the gradual buildup of matter, absorbing disc material courtesy of its magnetic field. The planets will form through a similar process, and their type will depend on their location according to the so-called "frost line," the distance from the protostar where temperatures are cold enough for compounds such as water, ammonia, and methane to condense into solid ice. Bodies beyond the line solidify and grow, attracting large amounts of gas. Inside the line, high temperatures evaporate the lighter elements, leaving only rocks and metals to form small planetary seeds that grow by colliding with each other.

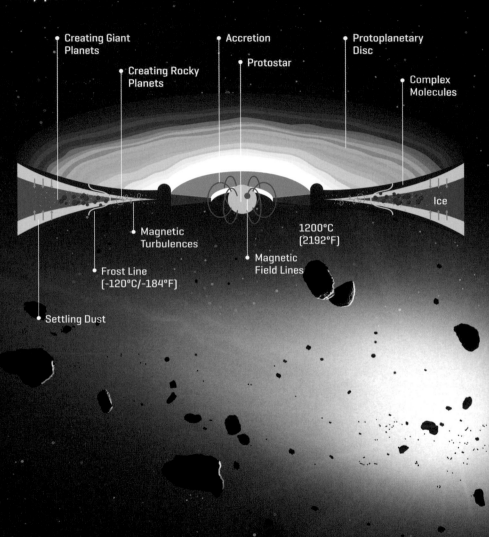

Creating Giant Planets

Creating Rocky Planets

Accretion

Protostar

Protoplanetary Disc

Complex Molecules

Magnetic Turbulences

Frost Line [-120°C/-184°F]

Magnetic Field Lines

1200°C [2192°F]

Ice

Settling Dust

The Dynamics of a Protoplanetary Collision

The rocky planets, including Earth, got their final form by colliding and fusing with other protoplanetary bodies of considerable mass.

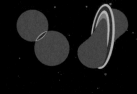

Two bodies of similar size collide, and the impact deforms them.

Heat makes the materials mix and collapse under the force of gravity.

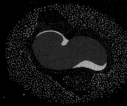

The iron cores move to the center of the combined mass.

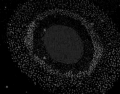

The new proto-Earth starts to rotate rapidly.

The debris floating around the rotating proto-Earth forms the moon.

A BURNING COLOSSUS
This rocky protoplanet very close to its star has a large enough mass to become spherical, even though its surface is continuously bombarded by asteroids.

MARS AT GROUND LEVEL
This self-portrait of the Curiosity rover, taken in the area known as Murray Buttes at the base of Mount Sharp, combines 60 photographs to give us a fuller picture of the planet's landscape. Its reddish color comes from iron oxides on its surface.

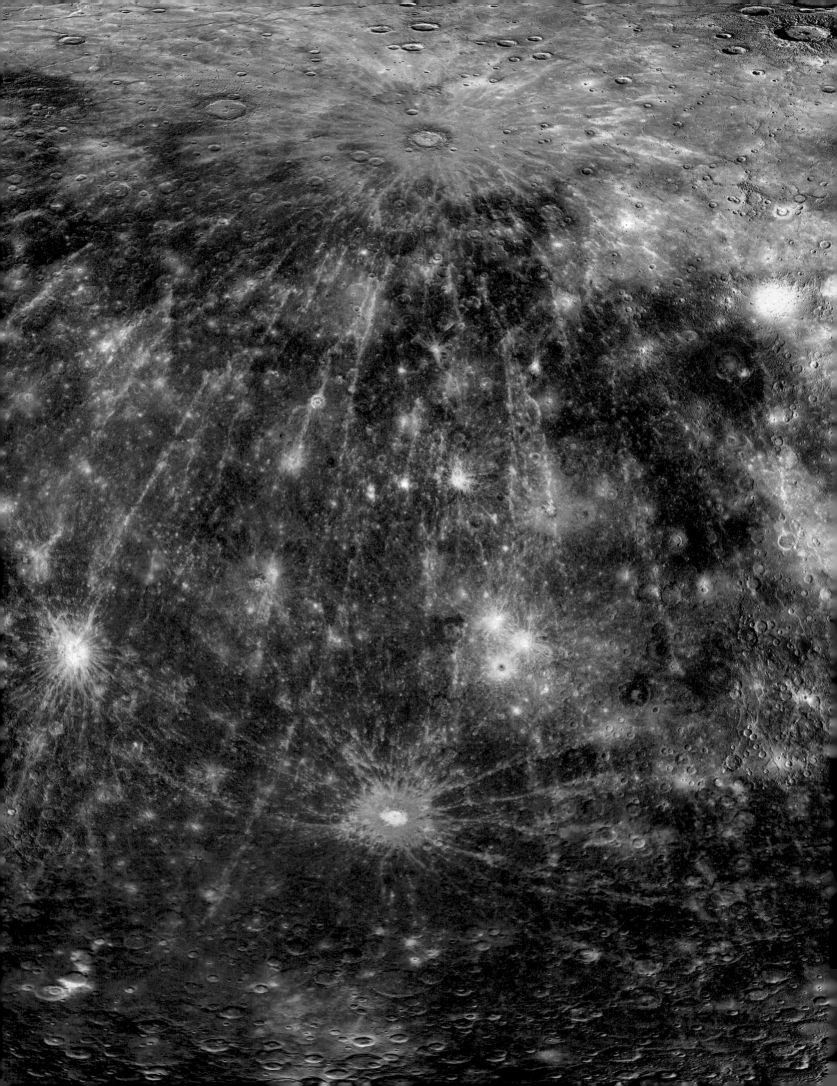

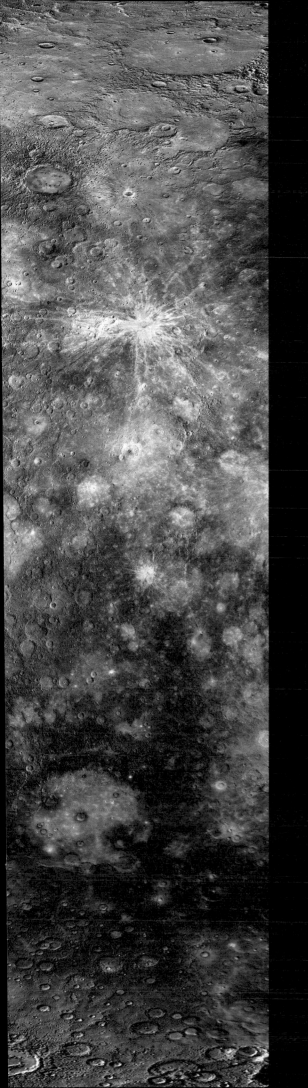

THE EARTH
AND OTHER
ROCKY PLANETS

This mosaic of the Mercurian surface comes from photos taken by the Messenger probe. The colors have been saturated to highlight the planet's surface, revealing more than the human eye can see.

WORLDS OF ROCK

Earth

1 Earth is the densest planet in the solar system, with an average density of 5.51 g/cm³, and also the largest of the rocky planets. Its core is made mostly of iron, nickel, and sulfur. The inner section is solid, while the external part is partially melted and has convection currents because of its magnetic field. The mantle is made of silicon and magnesium oxides, while the crust is composed of silicates. One of Earth's most distinctive characteristics is the presence of liquid water on the surface, averaging a temperature of 14°C (57°F).

Venus

2 Venus is similar to Earth in size and mass, which may mean its composition and internal structure are too. Its surface is covered by more volcanoes than any other planet, many of which can be explained by a long-ago large impact that would also have caused its retrograde rotation. Its temperature, which averages 460°C (860°F), is high, and it barely changes because of an intense greenhouse effect and strong winds that redistribute the heat.

Mars

3 Mars has 15 percent of Earth's volume and only 10.7 percent of its mass, making it less dense than our planet. In fact, its density is 3.94 g/cm³, which suggests a large number of light elements in its core. The surface of Mars is the most Earth-like—it even has geological formations like dunes (pictured above) and water in the form of both ice and water vapor. The average temperature is estimated at −63°C (−81.4°F).

Mercury

4 The smallest and least massive planet in the solar system, Mercury is the second densest after the Earth. The planet is more than 70 percent metallic elements, mostly partially molten iron, which suggests the presence of a magnetic field. A large number of impact craters shows that there is little geological activity. Its thin atmosphere and long rotation period (59 Earth days) cause a lot of variation in its temperature, which swings from a high of 430°C (806°F) to a low of −190°C (−310°F).

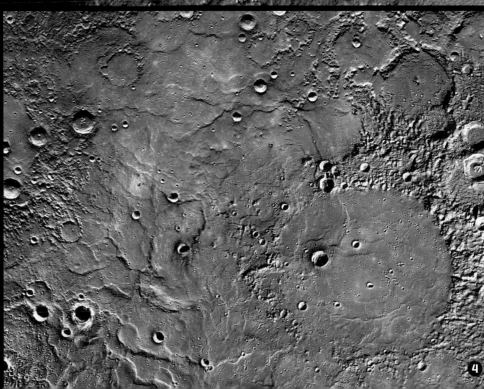

LIGHT ATMOSPHERES

The atmospheres of the rocky planets are thin in comparison to those of their gas giant counterparts. In Earth's case, its thin layer of gas is part of what allows for life.

The outer planets have enough mass to trap large amounts of most gases, including hydrogen and helium. The inner planets capture varying amounts, which results in different atmospheric compositions. Earth and Venus, for example, attract water vapor, methane, and ammonia, but Mars does not. There are other bodies in the solar system with atmospheres worth mentioning, including Titan, one of Saturn's many moons, and Pluto.

Protector of Life

The atmosphere plays an important role in protecting life on Earth from the most dangerous solar radiation. It consists mostly of nitrogen and oxygen, with trace amounts of other gases. Water vapor and carbon dioxide are crucial for storing the sun's energy in the form of heat.

BLUE SUNSET, RED PLANET
The Martian sunset has a uniform blue tint because of how the planet's dust scatters light. Blue light is scattered less than other colors, so it hangs around the sun. Because there are no clouds, there are no differences in intensity.

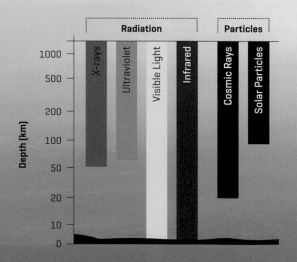

A SHIELD AGAINST RADIATION

Solar energy reaches Earth in the form of electromagnetic radiation and charged particles, which are absorbed by the atmosphere before and after they bounce off the surface. Ionized radiation (x-rays, ultraviolet rays, and cosmic rays) can change atomic and molecular properties, which is harmful to life but does not travel farther than about 20 kilometers (12 mi) into the atmosphere.

Chart — Depth [km] vs. Radiation and Particles:

	Radiation				Particles	
1000	X-rays	Ultraviolet	Visible Light	Infrared	Cosmic Rays	Solar Particles
500						
200						
100						
50						
20						
10						
0						

Mars's Atmosphere

The Martian atmosphere consists primarily of carbon dioxide, nitrogen, and traces of water vapor, and is very thin: the pressure on its surface is about 0.6 percent that of Earth's mean sea level pressure.

Venus's Atmosphere

Its high density makes its pressure 90 times that of Earth's. Clouds of sulfuric acid reflect most of the sun's energy, but an intense greenhouse effect makes its surface the hottest in the solar system.

Earth's Atmosphere

The troposphere, the layer closest to the surface, is where most meteorological phenomena take place. Above it, the stratosphere contains the ozone layer, which filters ultraviolet radiation. The atmosphere's temperature depends on how far you are from the surface, but it increases in the thermosphere due to the absorption of high-frequency solar radiation. The exosphere melts away into space.

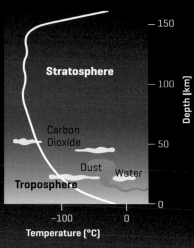

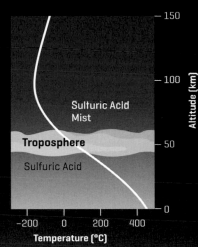

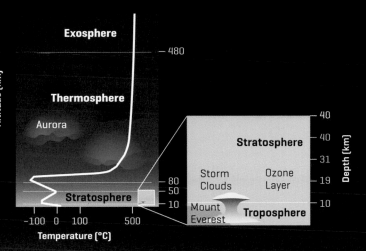

NIGHT LIGHTS
Noctilucent, or night-shining, clouds are made of ice and meteor debris. Although their surface is in darkness, they form in the mesosphere at an altitude that allows the last rays of the sun to illuminate them.

MERCURY, THE CLOSEST PLANET TO THE SUN

Mercury is the smallest and least massive planet in the solar system. Its proximity to the sun has made it difficult to study, but researchers have managed to obtain information about its composition and atmosphere.

Though Mercury is even smaller than some moons in the solar system, it is dense because of its large iron core. Its mantle and crust, in contrast, are relatively thin. According to the most accepted theory, it developed this unusual structure when it lost its mantle after an impact with a planetesimal during the formation of the solar system. At first glance, the Mercurian landscape is very similar to the moon, its surface riddled with impact craters created by celestial bodies as the planet was forming. Some of them once had rivers of lava, a result of volcanic activity that stopped some 750 million years ago, that smoothed patches of the surface. The most notable crater is the Caloris Planitia, formed by a violent impact that caused fractures and escarpments (abrupt slopes) on the opposite side of the planet.

A Tenuous and Unstable Atmosphere

Because of its small size and high temperatures (over 430°C/800°F in the regions facing the sun), Mercury cannot maintain a stable atmosphere. Instead it is covered by a very thin gaseous layer known as an exosphere. The atmospheric pressure on the surface is only 10 to 11 millibars (mbar) compared with the 1,013 mbar found on Earth at sea level.

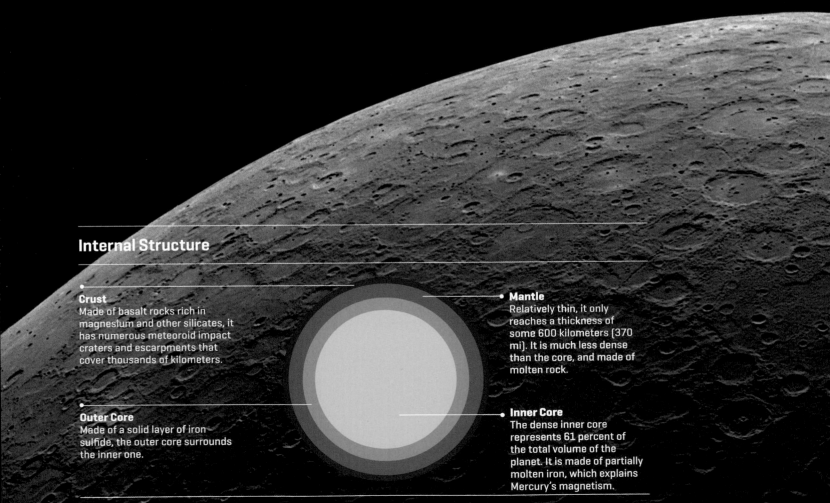

Internal Structure

Crust
Made of basalt rocks rich in magnesium and other silicates, it has numerous meteoroid impact craters and escarpments that cover thousands of kilometers.

Mantle
Relatively thin, it only reaches a thickness of some 600 kilometers (370 mi). It is much less dense than the core, and made of molten rock.

Outer Core
Made of a solid layer of iron sulfide, the outer core surrounds the inner one.

Inner Core
The dense inner core represents 61 percent of the total volume of the planet. It is made of partially molten iron, which explains Mercury's magnetism.

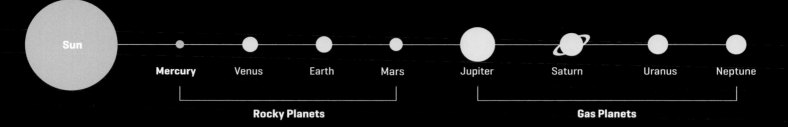

Sun

Mercury   Venus   Earth   Mars   Jupiter   Saturn   Uranus   Neptune

Rocky Planets

Gas Planets

CALORIS PLANITIA

With a diameter of 1,550 kilometers [963 mi], the Caloris Planitia [pictured as a large yellow spot] is the most prominent feature on Mercury's surface and one of the largest impact craters in the solar system. The energy from the impact threw material 1,000 kilometers [621mi] from the crater's border. Lava rose through the mantle, creating a ring nearly 2 kilometers [1.2 mi] high.

EXOSPHERE

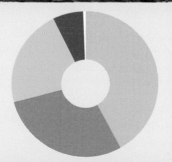

- **Oxygen** 42%
- **Sodium** 29%
- **Hydrogen** 22%
- **Helium** 6%
- **Potassium** 0.5%
- **Other components** [argon, carbon dioxide, water, nitrogen, xenon, krypton, neon, calcium, magnesium] 0.5%

MERCURY AT A DISTANCE
This view of Mercury, which comes from a mosaic of images captured by the Messenger space probe, shows the pock-marked details on the surface of the planet.

VENUS, THE HOTTEST PLANET

Venus is Earth's closest planetary neighbor and the most similar in terms of size and composition, but that's where the similarities end. On this planet molded by volcanic activity, a dense atmosphere traps solar radiation, bringing it hellish temperatures.

Venus is similar to Earth in many ways: in size, mass, and composition. It might even have been habitable at one point in its life. However, today's Venus is far from a paradise. With an atmosphere of mostly carbon dioxide, Venus is covered by a dense layer of sulfuric acid clouds that trap radiation in a severe greenhouse effect, making it the hottest planet in the solar system, with an average temperature of 464°C (867°F). The atmosphere is so dense that the surface pressure is 90 times that of Earth's.

Landscapes Molded by Volcanoes

Despite its extreme atmospheric conditions, Venus was the first neighboring planet we explored with a space probe. The Soviet Venera 7 ventured out in 1970, sending images back to Earth from its surface. The photographs showed a landscape marked by intense past volcanic activity, and even some evidence that active volcanoes may still exist. Most of the planet's surface appears to have formed rather recently (a few hundreds of millions of years ago), which

suggests that Venus may have suffered a cataclysmic event that caused its crust to reform. A curious aspect of our extreme neighbor is its rotation: it turns in the opposite direction from all other planets in the solar system besides Uranus. We don't yet understand what caused its retrograde rotation, but it could also be due to that same cataclysm.

THE VENUSIAN SURFACE
The planet's surface is relatively young—only 400 million years old—and features plains that host crisscrossing rivers of lava, volcanoes, and mountains. This image was created from the data transmitted by the Magellan probe.

A HABITABLE WORLD

The surface of Venus is one of the most hostile environments we've ever encountered. But was it always like this? From the data obtained from the Magellan probe, scientists have re-created the planet's climate at various points in its past. The results indicate that Venus could have had an average surface temperature of 11°C (52°F) some 3 billion years ago, only increasing some 4°C (39°F) over the following 1.2 billion years due to the sun's evolution. This illustration shows a land–sea pattern that could have made the planet habitable.

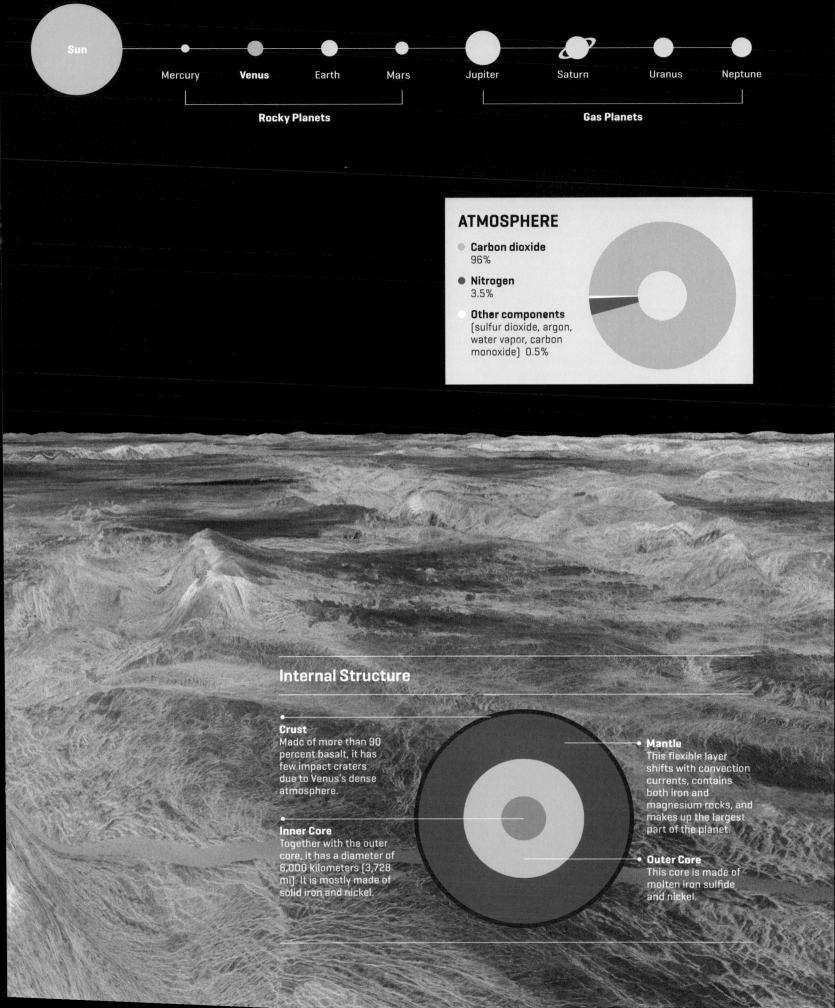

Sun

Mercury **Venus** Earth Mars Jupiter Saturn Uranus Neptune

Rocky Planets **Gas Planets**

ATMOSPHERE

- **Carbon dioxide**
  96%

- **Nitrogen**
  3.5%

- **Other components**
  (sulfur dioxide, argon,
  water vapor, carbon
  monoxide) 0.5%

Internal Structure

Crust
Made of more than 90
percent basalt, it has
few impact craters
due to Venus's dense
atmosphere.

Inner Core
Together with the outer
core, it has a diameter of
6,000 kilometers (3,728
mi). It is mostly made of
solid iron and nickel.

Mantle
This flexible layer
shifts with convection
currents, contains
both iron and
magnesium rocks, and
makes up the largest
part of the planet.

Outer Core
This core is made of
molten iron sulfide
and nickel.

EARTH, THE BLUE PLANET

Its size and orbital distance from the sun allow the largest
of the rocky planets to maintain a protective atmosphere and liquid
water—conditions that favor life.

Slightly larger than Venus, Earth is the third closest planet to the sun. Although fifth in terms of size, it has the highest density of all the planets in the solar system.

Unique Characteristics
Earth enjoys unique conditions both inside and out. It sits in the habitable zone of the solar system, which means the sun's radiation allows for water to exist as a liquid on the surface. The temperature on Earth's surface is very close to water's "triple point," which is when its three basic states—solid, liquid, and gas—can exist.

In fact, water occupies three-fourths of Earth's surface. Its atmosphere is made up mostly of nitrogen and oxygen, which are important for biological activity, and our ozone layer and magnetic field protect against the most energetic solar particles. The lithosphere around the planet's outer layers (the crust and upper mantle) is divided into plates that are constantly moving, causing earthquakes and volcanic activity that in turn form mountains and ocean trenches. Earth has one natural satellite, our moon, which is the largest in the solar system in relation to its planet.

ATMOSPHERE

- **Nitrogen** 78%
- **Oxygen** 21%
- **Argon** 0.9%
- **Other gases**
  (carbon dioxide, neon, helium, methane) 0.1%

Internal Structure

Crust
The oceanic crust, made of basalt rocks, is different from the continental crust, which is thicker and made mostly of granite rocks. Combined with the mantle they form the lithosphere, which is divided into tectonic plates: enormous rocky sheets that move because of the convection currents in the mantle.

Mantle
The mantle is the largest of Earth's layers, accounting for 84 percent of its volume. It is made of silicate rocks rich in iron and magnesium.

Outer Core
Molten iron and nickel make up the outer core, which creates Earth's magnetic field.

Inner Core
The inner core is solid and made of an iron–nickel alloy.

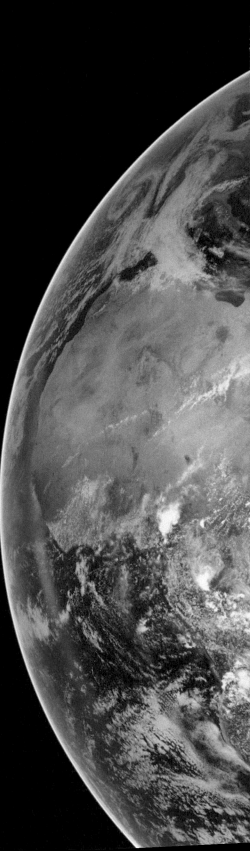

Sun

Mercury   Venus   **Earth**   Mars   Jupiter   Saturn   Uranus   Neptune

Rocky Planets   **Gas Planets**

BLUE SPHERE
This image of Earth shows
how the oceans reflect the
color of the atmosphere,
whose composition makes
blue wavelengths more
visible than other colors.

THE EARTH AND THE MOON

Earth's moon is the largest in the solar system in relation to the planet it orbits. We think it was born some 4.5 billion years ago, formed from debris thrown up when the proto-Earth collided with a hypothetical planet with dimensions similar to Mars.

At more than a quarter (27 percent) the size of Earth, the moon is almost large enough to be considered our twin planet. The moon's visible surface covers more than a quarter the area of Earth's continents. The rugged surface, riddled with impact craters, is proof of its lack of both an atmosphere and tectonic activity. Because it has no gaseous, protective layer, its surface temperatures range from 100°C (212°F) during the day to –150°C (–238°F) at night. Recorded maximum and minimum temperatures have reached 122.7°C (252.9°F) and –232°C (–386°F), making the moon one of the coldest places in the solar system.

Collateral Planet

The moon likely formed when an infant Earth collided with a hypothetical planet during the initial phases of its creation, hurling out debris that melded into a molten mass and eventually cooled. This theory comes from the fact that the moon and Earth rocks have identical proportions of chemical isotopes, which indicates a common origin, but since the moon's density is lower and its nucleus has very little iron, the two couldn't have formed at the same time.

A Dark World

The moon may seem bright, but it only reflects around 3 to 12 percent of the sunlight it receives as opposed to the approximately 30 percent Earth reflects. This dampening is because it lacks transition metals.

Mutual Attraction

The moon both orbits the Earth and turns on its axis every 27 days—a symmetry that keeps the moon tidally locked, which means we only ever see one of its faces. The moon's size and proximity to Earth encourage gravitational interaction between the two bodies. That push and pull is responsible for both Earth's ocean tides and earthquakes on the moon.

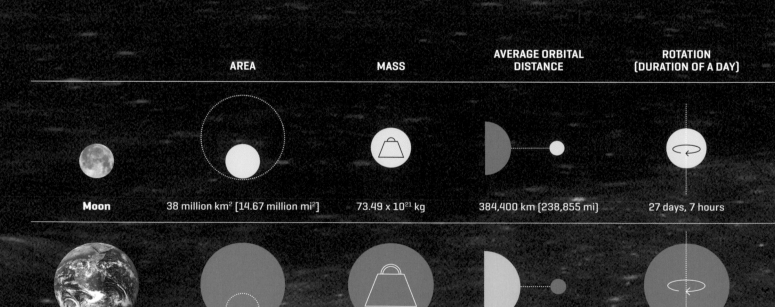

	AREA	MASS	AVERAGE ORBITAL DISTANCE	ROTATION (DURATION OF A DAY)
Moon	38 million km² [14.67 million mi²]	73.49 x 10²¹ kg	384,400 km [238,855 mi]	27 days, 7 hours
Earth	510.1 million km² [196.95 million mi²]	5.972 x 10²⁴ kg	150 million km [93,206 mi]	23 h 56 min

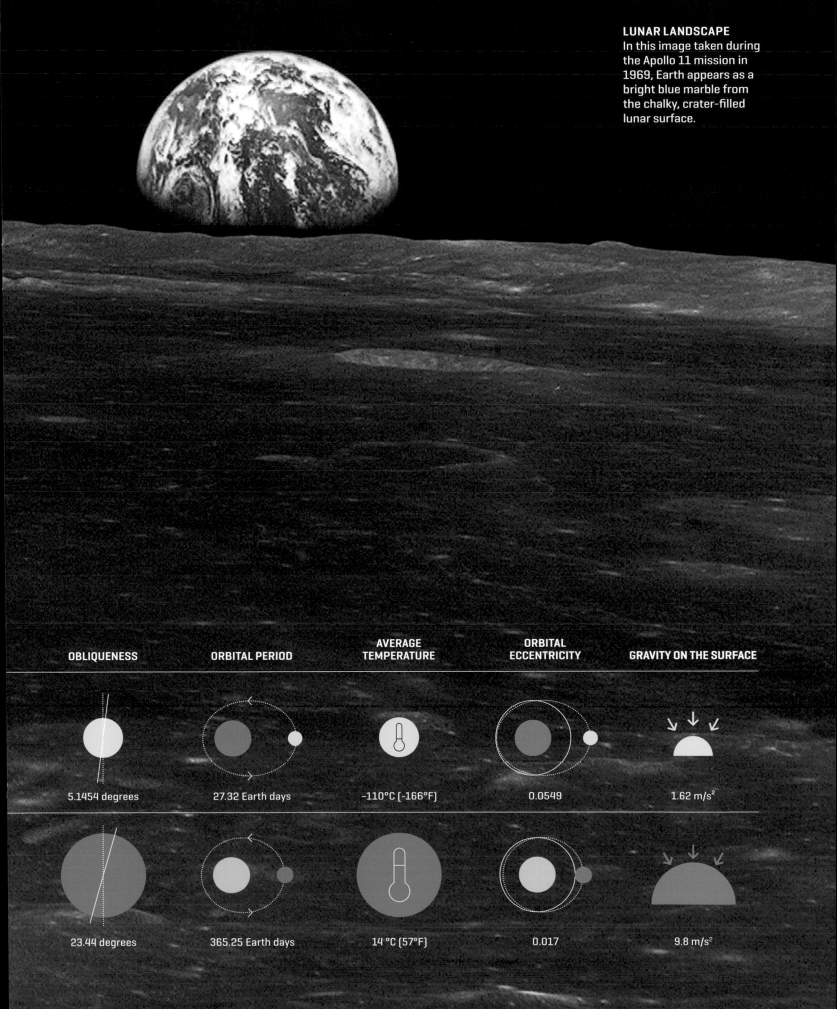

OBLIQUENESS	ORBITAL PERIOD	AVERAGE TEMPERATURE	ORBITAL ECCENTRICITY	GRAVITY ON THE SURFACE
5.1454 degrees	27.32 Earth days	−110°C (−166°F)	0.0549	1.62 m/s²
23.44 degrees	365.25 Earth days	14 °C (57°F)	0.017	9.8 m/s²

MARS, THE LAST ROCKY PLANET

Mars has a little more than half of Earth's diameter, but it has an internal structure similar to our planet and a surface not unlike some of Earth's driest places.

Mars is the fourth rocky planet from the sun. Beyond it lies the asteroid belt, the region between Mars and Jupiter where most of the solar system's asteroids are found. The second closest planet to Earth has some characteristics in common with it, encouraging scientists to further explore it. The Red Planet gets its color from an abundance of iron oxide. Its atmosphere is thin and made up of carbon dioxide, and its surface pressure is only 7 mbar (some 1,000 times less than Earth's). The weak atmosphere and lack of a global magnetic field mean that Mars has no effective shield against radiation and solar wind. Temperatures range from –87ºC (–125ºF) to 20ºC (68ºF).

Mars at Ground Level

The surface of Mars looks almost like the moon, except that it has polar caps, tall volcanoes, and enormous canyons.

Most of its surface was formed by magma (molten rock) from the mantle, an interior layer that was liquid in the first stages of the planet's evolution. The Red Planet is still geologically active in places: its polar caps show seasonal changes, the dune fields shift in its winds, and small streams seem to appear seasonally. In fact, the European Space Agency's Mars Express, first launched in 2003 and expected to be active through 2020, discovered a saltwater lake under the caps of ice at the south pole. Some of Mars's conditions and landscapes remind us of those found on Earth. Some 3.8 billion years ago, Mars had a thicker atmosphere, a magnetic field induced by the rotation of its nucleus, a hotter climate, and large oceans of liquid water, so it was possible that the planet held some kind of life. But the conditions changed, turning Mars into the mostly dry, cold planet we know today.

Internal Structure

Crust
Basalt rock and iron oxide are the main components of Mars's crust. Iron oxide is what gives Mars its distinctive red color.

Mantle
Millions of years ago, Mars's mantle was liquid and its magma sculpted the crust's surface, which is made of iron oxide and olivine.

Core
The core is made of solid iron, nickel, and 16 percent sulfur. Its small size and slow rotation created a weak magnetic field.

ATMOSPHERE

- Carbon dioxide 95.3%
- Nitrogen 2.7%
- Argon 1.6%
- Other gases (oxygen, water vapor, carbon monoxide) 0.4%

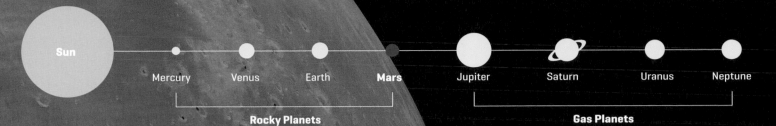

Sun

Mercury   Venus   Earth   **Mars**   Jupiter   Saturn   Uranus   Neptune

Rocky Planets

Gas Planets

THE METHANE MYSTERY

On Earth, methane is a gas produced by living beings, so it was a surprise when the Mars Express mission found it in the Martian atmosphere in 2004. Thanks to the spectrometers of the Mauna Kea telescopes in Hawaii, scientists put together the first map of methane on Mars (pictured below). The planet is estimated to produce some 150 tons of methane per year, although we still don't know how it forms and disappears. In 2020 the joint European Space Agency/ Roscosmos ExoMars mission will tackle the mystery, creating a precise map of its methane distribution with the hope of determining its origin.

VALLES MARINERIS
Made with 102 images taken by the Viking 1 orbiter on February 22, 1980, this mosaic of Mars shows the majestic Valles Marineris at its center: a system of canyons 4,000 kilometers (2,500 mi) long and 7 kilometers (4 mi) deep.

Concentration of methane (parts per billion)

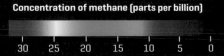

30   25   20   15   10   5   0

MARS AND EARTH

Though it's smaller and lighter than Earth, and not even our closest neighbor, Mars is considered a kind of sister planet whose characteristics allow for the possibility of future colonization.

Mars is the solar system's most Earth-like planet. The length of its days and the incline of its axis of rotation are almost identical to Earth's; its seasons, polar caps, and clouds are similar too. The main difference between the two is the average surface temperature: Mars is colder. The Red Planet's lack of liquid water and its thicker atmosphere also make it different.

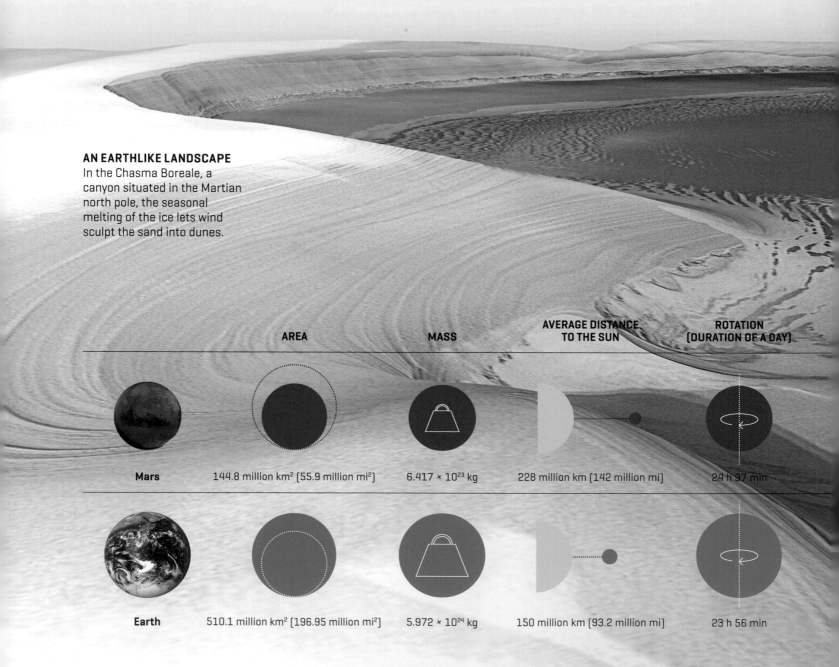

AN EARTHLIKE LANDSCAPE
In the Chasma Boreale, a canyon situated in the Martian north pole, the seasonal melting of the ice lets wind sculpt the sand into dunes.

	AREA	MASS	AVERAGE DISTANCE TO THE SUN	ROTATION (DURATION OF A DAY)
Mars	144.8 million km² (55.9 million mi²)	6.417×10^{23} kg	228 million km (142 million mi)	24 h 37 min
Earth	510.1 million km² (196.95 million mi²)	5.972×10^{24} kg	150 million km (93.2 million mi)	23 h 56 min

The Cycles of the Seasons

The sun's north pole is in the center of the graphic at right, which shows a top-down view of the solar system. The outer ring of each orbit (Mars and Earth) shows the seasons in the northern hemisphere, while the interior ring shows the seasons in the southern hemisphere. Mars and Earth have a similar obliqueness, which means their seasons almost align. Except the Martian year is approximately twice as long as Earth's, which means its seasons are too. Given that Mars's orbit is more elliptical, some of its seasons are longer than others: the northern hemisphere has a longer spring and summer than the southern hemisphere, which has a longer fall and winter.

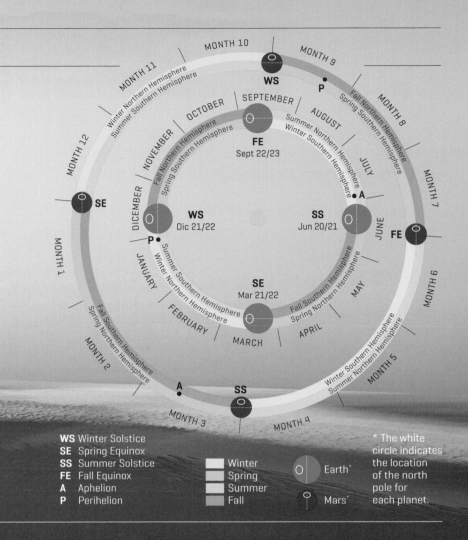

Northern Hemisphere Season	Length in Earth Days	
	Earth	Mars
Spring	93	194
Summer	93	178
Fall	90	142
Winter	89	154

WS Winter Solstice
SE Spring Equinox
SS Summer Solstice
FE Fall Equinox
A Aphelion
P Perihelion

Winter
Spring
Summer
Fall

Earth*

Mars*

* The white circle indicates the location of the north pole for each planet.

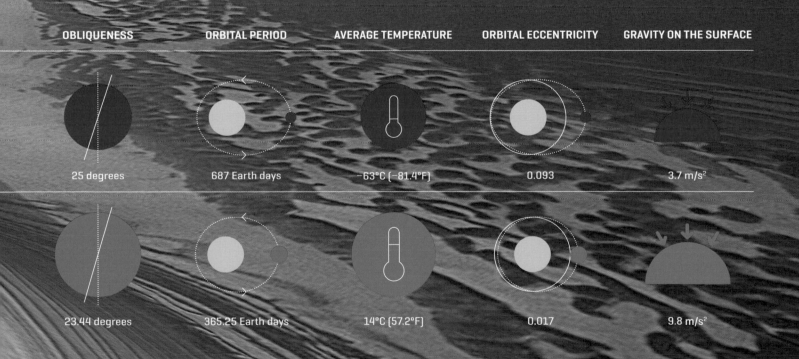

	OBLIQUENESS	ORBITAL PERIOD	AVERAGE TEMPERATURE	ORBITAL ECCENTRICITY	GRAVITY ON THE SURFACE
Mars	25 degrees	687 Earth days	−63°C (−81.4°F)	0.093	3.7 m/s²
Earth	23.44 degrees	365.25 Earth days	14°C (57.2°F)	0.017	9.8 m/s²

MAP OF MARS

Mars's geology has many notable and intriguing aspects, among them the twisting line of topography drawing a boundary between ancient, rugged ground (warm colors) and more modern, flat regions (cool colors).

The origin of Mars's north/south divide, also known as the Martian dichotomy, has been the subject of numerous debates. Scientists have many theories about what created it: a large impact, catastrophic internal processes, and Martian tectonic plates, to name a few. The one thing we do know is that the surface of the southern two-thirds has a large number of craters, while the upper third has mostly flat plains and is much less rugged.

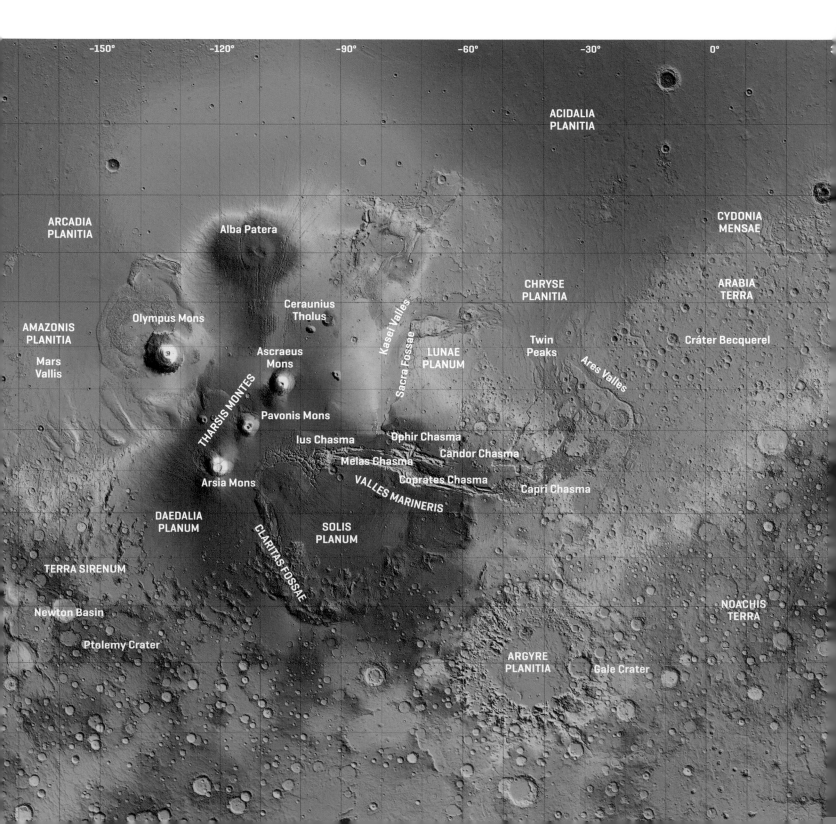

TOWERING VOLCANOES

Mars's volcanoes are colossal and among the highest peaks in the solar system.

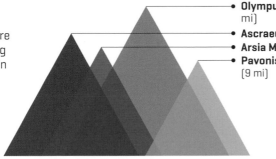

- **Olympus Mons** 22 km (14 mi)
- **Ascraeus Mons** 18 km (11 mi)
- **Arsia Mons** 16 km (10 mi)
- **Pavonis Mons** 14 km (9 mi)

TOPOGRAPHICAL MAP OF MARS

The Mars Orbital Laser Altimeter gathered the data for the global topographical map of the red planet below, with the altitudes of the two hemispheres indicated by color.

Height in km

-8 -4 0 4 8 12

MALEA PLANUM

UTOPIA PLANITIA

Hecates Tholus

Elysium Mons

Albor Tholus

Cerberus Fossae

Orcus Patera

Nili Fossae

ISIDIS PLANITIA

SYRTIS MAJOR PLANUM

Aeolis Mons

Apollinaris Patera

Gusev Crater

TERRA CIMMERIA

Dao Vallis and Niger Vallis

HELLAS PLANITIA

MALEA PLANUM

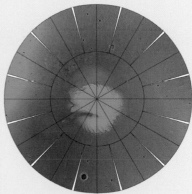

Latitude: from 60° to 90° N

NORTHERN HEMISPHERE

This hemisphere was formed by more recent geological activity and is relatively flat with few craters. The crust reaches a maximum thickness of 32 kilometers (20 mi).

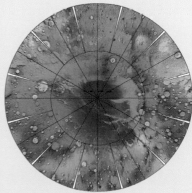

Latitude: from 60° to 90° S

SOUTHERN HEMISPHERE

The older grounds to the south take up two-thirds of the Martian surface and date from the time of a great meteor bombardment. The crust reaches a thickness of 58 kilometers (36 mi).

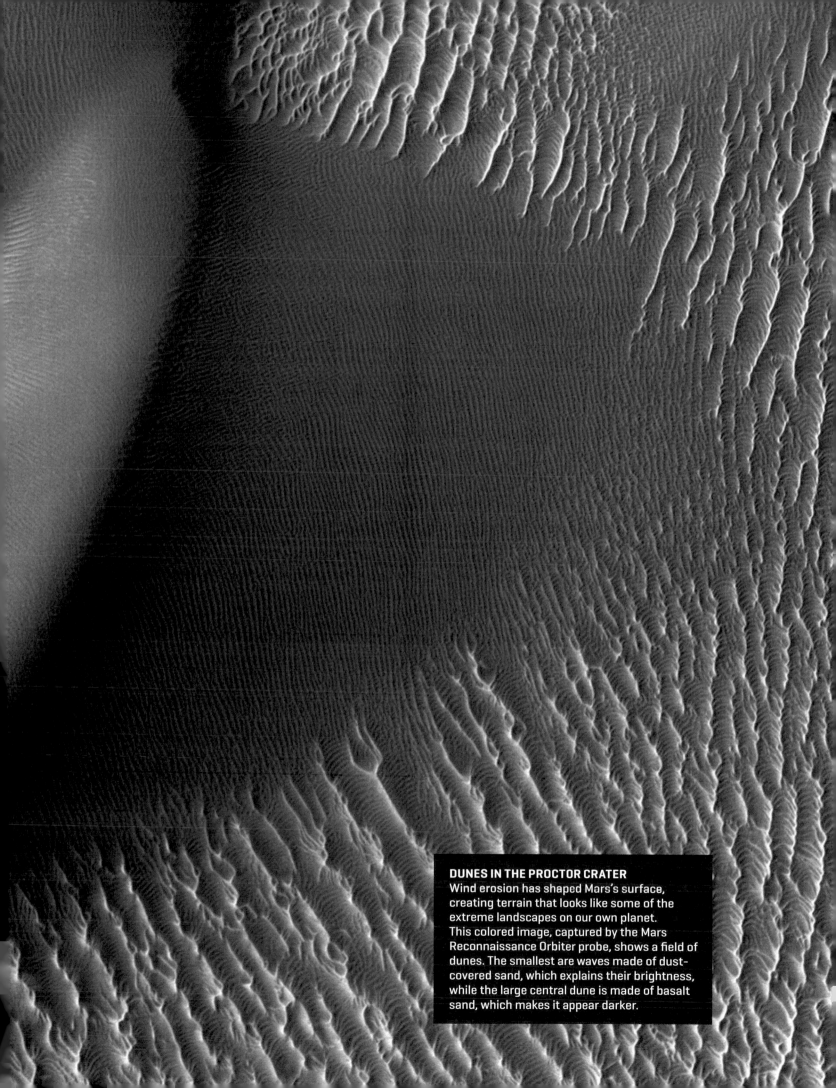

DUNES IN THE PROCTOR CRATER
Wind erosion has shaped Mars's surface, creating terrain that looks like some of the extreme landscapes on our own planet. This colored image, captured by the Mars Reconnaissance Orbiter probe, shows a field of dunes. The smallest are waves made of dust-covered sand, which explains their brightness, while the large central dune is made of basalt sand, which makes it appear darker.

VOLCANOES IN THE SOLAR SYSTEM

All rocky planets, not just Earth, show signs of volcanic activity, whether it's ongoing or long past. No such activity has been detected in the outer solar system, with one notable exception: Jupiter's moon Io.

Most of the rocky planets' surfaces are covered with volcanoes—places where subterranean magma is sometimes ejected—though most are currently inactive. The characteristics of volcanic activity differ from planet to planet.

Mars's Volcanoes

Because of its lower gravity, Martian volcanoes have enormous eruptive force. Their lava flows were long-lasting and large, resulting in some of the biggest known volcanoes. The largest of those, Olympus Mons in the Tharsis volcanic region, is 22 kilometers (14 mi) tall. It is currently inactive, like most volcanoes on Mars.

Earth's Volcanoes

On Earth, volcanic activity occurs along the edges of tectonic plates as they grind against each other. This activity is not limited to volcanoes: in fact, large amounts of lava and gas emissions occur along midocean ridges.

Volcanoes on Other Bodies in the Solar System

Venus has more than 1,600 volcanoes, but the planet's dense atmosphere makes them difficult to observe. Venus does not have tectonic plates, and its crust does not slide, but rather rises and sinks. In 2015 the Venus Express probe found indications of volcanic activity.

Volcanoes in the Outer System

Among Jupiter's largest moons, Io is the closest one to it. Its proximity is the main reason for its geological activity, which is the most intense in the solar system. The constant gravitational push and pull from the huge gas giant and neighboring moons creates Io's huge tidal forces. The resulting friction heats the moon up, causing volcanic activity. Io has more than 400 active volcanoes, some of which produce clouds of sulfur and sulfur dioxide that rise several hundred kilometers.

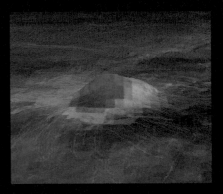

Venusian Hot Zone
This image of the Venusian volcano Idunn Mons has been color coded to show the hottest zones in red. The top of the volcano is significantly hotter than its surroundings, which suggests a recent lava emission.

A DWELLING FIT FOR THE GODS
Crowned by an impressive caldera (a bowl-shaped pit created by a volcano's collapse) and standing 22 kilometers (14 mi) tall, Olympus Mons dominates the Martian region of Tharsis. For its majesty it was named after the home of the ancient Greek gods.

A WORLD OF SULFUR
Sulfur and sulfur dioxide, lava, and ejections from hundreds of active volcanoes combine to give Io its characteristic brown-yellow aspect spotted with black and red.

A Sleeping Giant

Three times higher than Mount Everest and the size of the state of Arizona, Olympus Mons is the volcanic king of the solar system. This 30-million-year-old giant has a maximum diameter of 648 kilometers (over 400 mi). But Mars has other notable volcanoes. Massive Alba Mons is the largest of Mars's volcanoes by area: its surface area is comparable in size to that of the United States.

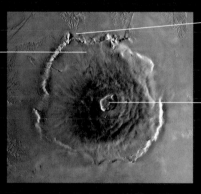

Halo
This vast plain stretches over hundreds of kilometers to the north and west of the summit.

Scarp
This line of cliffs surrounds the halo and reaches 6 kilometers (4 mi) high.

Caldera
Olympus Mons' caldera reaches up to 52 kilometers (32 mi) wide.

- Olympus Mons on Mars
- Mount Everest

VOLCANIC ERUPTION ON IO

This 1997 image from the Galileo orbiter shows a volcanic plume (in blue) on Jupiter's moon Io, shooting up some 100 kilometers (60 mi) into space. Its unique, umbrella-like shape is due to the moon's lack of an atmosphere combined with low gravity.

IO'S VOLCANOES
This illustration shows how Io's relentless eruptive activity has riddled its surface with conical volcanoes, calderas, and lava streams. Sulfur dyes the landscape yellow, and over the horizon the reddish aura is the result of light reflecting off suspended volcanic dust. This moon's lack of an atmosphere, plus a gravity five times weaker than Earth's, allow these volcanic plumes to eject gas and ash hundreds of kilometers upward.

THE GAS PLANETS, GIANTS OF THE SOLAR SYSTEM

Marbled Jupiter and its satellite Io, as seen from the New Horizons probe.
The planet has such a strong gravitational pull that it has deformed its moon,
making it the solar system's most active geological object in the process.

WORLDS OF GAS

The outer planets' cores are surrounded by enormous envelopes of gas and liquid that give them a size and mass much greater than those of their companions in the inner solar system.

Jupiter

1 Titanic Jupiter has as much mass as 318 Earths. Like all gas giants, it has a powerful magnetic field and a very thick atmosphere—it's basically a hydrogen ball around a solid core. It's made up mostly of hydrogen, which changes form depending on its distance from the core: it could be liquid and metallic deep inside the planet, while it takes the form of gas at the surface. The colors of its swirling surface depend on variations in temperature and composition, and whether the gas is sinking or rising. Frequent storms churn on Jupiter, like the famous Great Red Spot. First observed in the 17th century, the spot was once the size of two Earths, but it is shrinking.

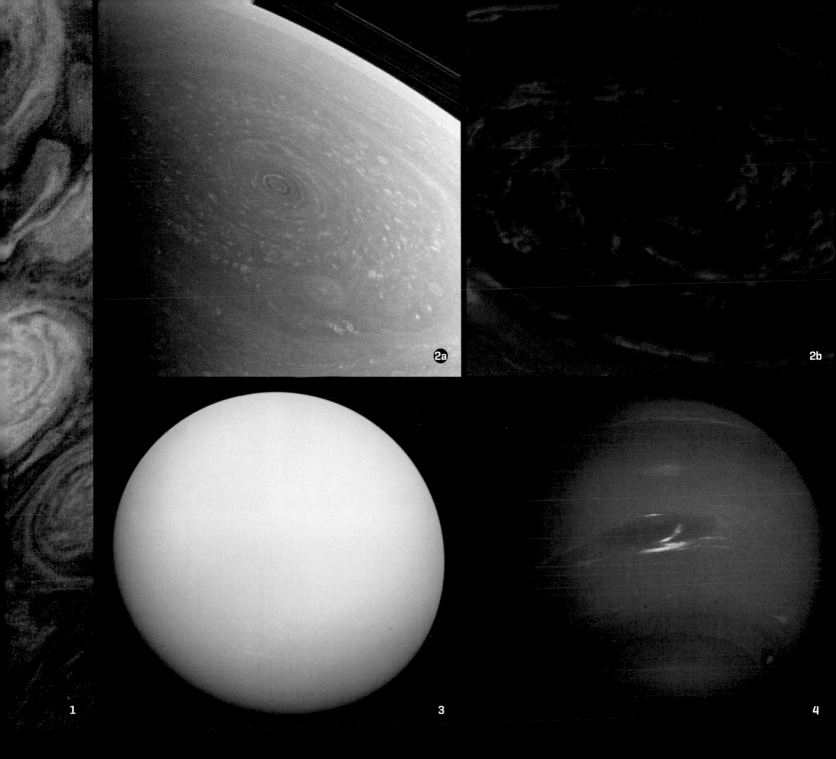

1

2a

2b

3

4

Saturn

2 It has the lowest density of all the planets: 0.7 g/cm³ (0.025 lb/in³). Its internal structure, powerful magnetosphere, and poles closely resemble those of its neighbor Jupiter, but because it has a lower mass and internal pressure its outer gas layer is probably thicker. This layer has bands of high and low pressure, like Jupiter, and storms that last hundreds of days (image 2b). The giant hexagon around the north pole (image 2a), whose center has a stormy vortex, is caused by jet streams in the atmosphere.

Uranus

3 Its very small core, which has only half the mass of Earth's, is metallic and rocky. The dense mantle around it is composed of water, methane, and ammonia ice with high electrical conductivity, while its outer gas envelope is made of hydrogen and helium. The planet's characteristic pale blue color could be due to methane. Although Voyager 2 did not detect much meteorological activity on Uranus, recent records indicate that it has more intense, windy zones than Jupiter. Uranus's most notable feature is its axis of rotation: 97.77 degrees. It rotates almost completely on its side.

Neptune

4 Its structure is very similar to that of Uranus, but Neptune's mantle is slightly larger and its gaseous layer is smaller. As the planet farthest from the sun, it receives the least amount of solar radiation: less than half that of Uranus. Even so, its atmospheric temperatures are similar, which reveals a source of internal heat that probably comes from material left over from its formation. This explains why its atmosphere is so dynamic, with winds that can reach 2,000 kilometers (over 1,240 mi) per hour and are the strongest in the solar system.

DENSE ATMOSPHERES

Because of their mass, Jupiter, Saturn, Uranus, and Neptune can capture gases such as hydrogen and helium with their gravitational pull. Their resulting atmospheres contain hurricane-force winds and electrical storms of an unimaginable intensity when compared with what we experience on Earth.

The gas planets' atmospheres are made up mostly of hydrogen and helium. Jupiter's is one of the most closely studied, thanks to the Galileo mission. Since the gas giants lack a solid surface, scientists use a determined pressure (105 Pa) and the planets' size as a reference in estimating the depth of its atmosphere. All of their characteristic cloudy layers fall below this point.

The Lord of the Storms

Jupiter experiences strong convection currents that divide the cloudy layer into bands of distinct colors, temperatures, and heights. The dynamics of these currents cause electrical storms and winds that travel up to 500 kilometers (310 mi) per hour.

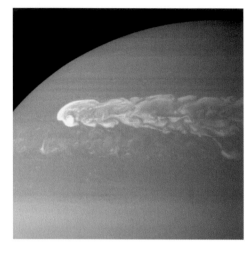

STORMS ON SATURN
A storm of unheard-of potential raged across the planet between 2010 and 2011.

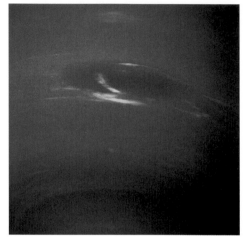

THE GREAT BLUE SPOT
Like Jupiter, Neptune has enormous spots on its surface: a sign of powerful anticyclones.

Polar Zones
Although less visible, the alternating bands of red clouds (bands) and white clouds (zones) go almost all the way to the poles.

Red Bands
The clouds that make up these bands are hot and low. Winds blow from west to east on the edge of the white zones.

White Zones
These layers of cold clouds are situated higher than the red ones, surrounded by westerly winds.

AURORAE ON SATURN
When the sun's charged
particles collide with
Saturn's magnetosphere, it
causes spectacular aurorae
at the planet's south pole.

The Composition of Atmospheres

Jupiter

The whole atmosphere is basically
made up of hydrogen. Clouds are found
at different levels according to their
composition: ammonia, ammonium
hydrosulfide, or water. The troposphere,
some 50 kilometers (31 mi) up, is very
turbulent and can have convective
storms that develop vertically up to 150
kilometers (93 mi).

Saturn

This planet's atmosphere and cloud cover
are similar to Jupiter's, although the clouds
are found higher up because of its lower
density. The Cassini probe, which orbited
Saturn from 2004 to 2017, detected the
largest storms in the solar system, with
lightning thousands of times stronger than
lightning on Earth.

Uranus and Neptune

They have the coldest atmospheres
in the solar system, earning them the
name "ice giants." One of their distinctive
characteristics is the presence of methane
clouds, which is what gives them their blue
tint. Neptune has the strongest winds in
the solar system: they can reach speeds of
2,000 kilometers (1,243 mi) per hour.

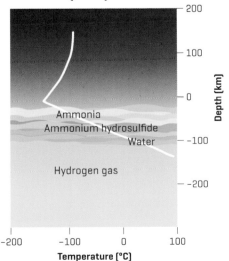

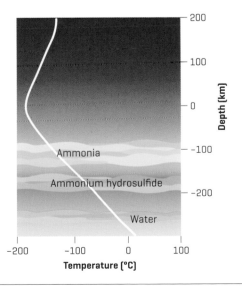

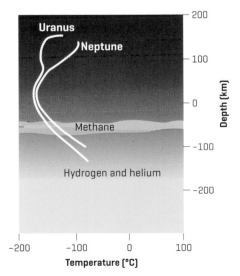

Jupiter is one of the brightest objects in Earth's night sky, which is why it was studied in ancient times by Mesopotamian astronomers. It is huge, with a mass 2.5 times the mass of the other seven planets combined.

Atmosphere and Magnetosphere

Jupiter's atmosphere is made mostly of hydrogen and helium. It has a distinctive look, divided into many colorful bands at different latitudes. It also features the Great Red Spot, a giant storm observed for the first time more than 300 years ago that is still venting its fury. Jupiter's rotation speed is slightly faster at its equator than at its polar axes, which means it is not a perfect sphere but rather a slightly flattened spheroid. Its interior could have a dense core made of a mix of different

elements surrounded by a layer of helium and metallic hydrogen (at very high pressures, hydrogen can become a liquid metal). This layer's electrical currents create the planet's intense magnetic field. Its magnetosphere is the largest and most powerful in the solar system, stretching out some 7 million kilometers (4.3 million mi) toward the sun and far enough to almost reach Saturn's orbit.

A Patchwork of Moons and Rings

Jupiter is surrounded by at least 79 moons, including those known as the Galilean moons: Io, Europa, Ganymede (the largest in the solar system), and Callisto. The gas giant also has its own ring system, though they are made of dust instead of ice and are not as spectacular as Saturn's.

Internal Structure

Atmosphere
The upper Jupiter's atmosphere are made mostly of hydrogen gas, with some helium.

Lower Atmosphere
Liquid hydrogen is found here due to pressure exerted by a layer of external clouds.

Layer Around the Nucleus
High temperature and pressure compress hydrogen atoms in this region into a layer of metallic liquid hydrogen.

Core Region
Rock, metal, and hydrogen compounds make up the region of Jupiter's solid core.

ATMOSPHERE

- **Hydrogen** 89%
- **Helium** 10%
- **Methane** 0.3%
- **Other gases**
  (ammonia, deuterium, ethane, water) 0.7%

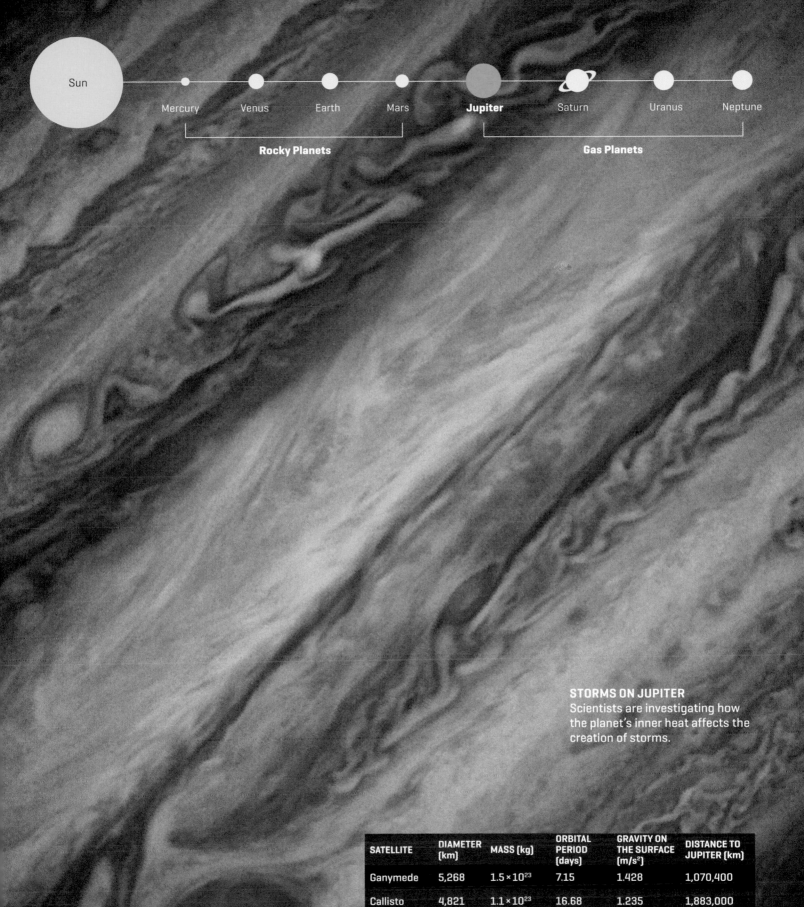

Sun

Mercury   Venus   Earth   Mars   **Jupiter**   Saturn   Uranus   Neptune

Rocky Planets   **Gas Planets**

STORMS ON JUPITER

Scientists are investigating how the planet's inner heat affects the creation of storms.

SATELLITE	DIAMETER (km)	MASS (kg)	ORBITAL PERIOD (days)	GRAVITY ON THE SURFACE (m/s²)	DISTANCE TO JUPITER (km)
Ganymede	5,268	1.5×10^{23}	7.15	1.428	1,070,400
Callisto	4,821	1.1×10^{23}	16.68	1.235	1,883,000
Io	3,642	8.9×10^{22}	1.76	1.796	421,700
Europa	3,122	4.8×10^{22}	3.55	1.314	607,900

JUPITER AND EARTH

Jupiter's measurements are overwhelming when compared to our little blue planet. It has a mass some 318 times greater than Earth's and a radius 11 times larger.

Its diameter, at 142,984 kilometers (88,846 mi), is 11 times wider than Earth's, and its volume, 1.4313×10^{15} km³, is 1,321 times greater. Its gravity is over 24.79 m/s², which is 2.5 times greater than our planet's (9.8 m/s²), requiring a much larger escape velocity to leave its gravitational attraction: 59.5 kilometers (37 mi) per second compared to 11.2 kilometers (7 mi) per second. But Earth surpasses Jupiter in other ways. For example, although it has the second highest density of all the gas giants (1.326 g/cm³/0.048 lb/in³), it is much lower than our planet's (5.514 g/cm³/0.199 lb/in³). Its distance from the sun means that Jupiter's orbital period is much longer than Earth's: it needs 11.86 years to complete one orbit. Furthermore, its orbital velocity is 13.07 km/s, much lower than our planet's, which reaches 29.78 kilometers (18.5 mi) per second.

Two Very Different Worlds

In the end, it is difficult to compare the two planets even though they are in the same neighborhood: the solar system. This gas giant's colossal dimensions, as well as its location and internal composition, make it an exotic and inhospitable place with surface conditions that are nothing like Earth's.

In Numbers

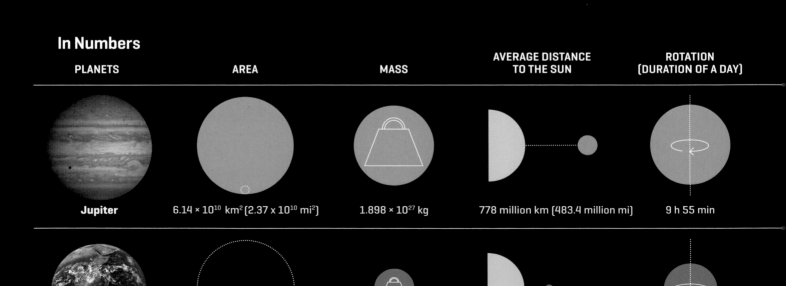

PLANETS	AREA	MASS	AVERAGE DISTANCE TO THE SUN	ROTATION (DURATION OF A DAY)
Jupiter	6.14×10^{10} km² (2.37×10^{10} mi²)	1.898×10^{27} kg	778 million km (483.4 million mi)	9 h 55 min
Earth	510.1 million km² (197 million mi²)	5.972×10^{24} kg	150 million km (93.2 million mi)	23 h 56 min

VIEWING JUPITER

Jupiter shines bright in this grouping of celestial bodies captured on location of the Very Large Telescope at the Paranal Observatory in Chile. Its brightness, which can reach a magnitude of -2.94 (only the moon and Venus are brighter, as shown here), is so intense that it can project shadows onto Earth's surface. These shadows are visible only under very specific conditions, unlike the moon's.

A GIANT PLANET

Everything on Jupiter is supersize, including the huge storm known as the Great Red Spot, seen here in its natural reddish-orange color. The storm is so big it could swallow Earth, and it used to be even larger.

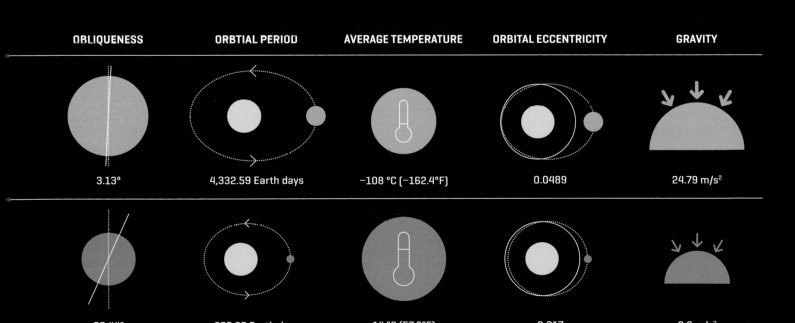

OBLIQUENESS	ORBTIAL PERIOD	AVERAGE TEMPERATURE	ORBITAL ECCENTRICITY	GRAVITY
3.13°	4,332.59 Earth days	−108 °C (−162.4°F)	0.0489	24.79 m/s²
23.44°	365.25 Earth days	14 °C (57.2°F)	0.017	9.8 m/s²

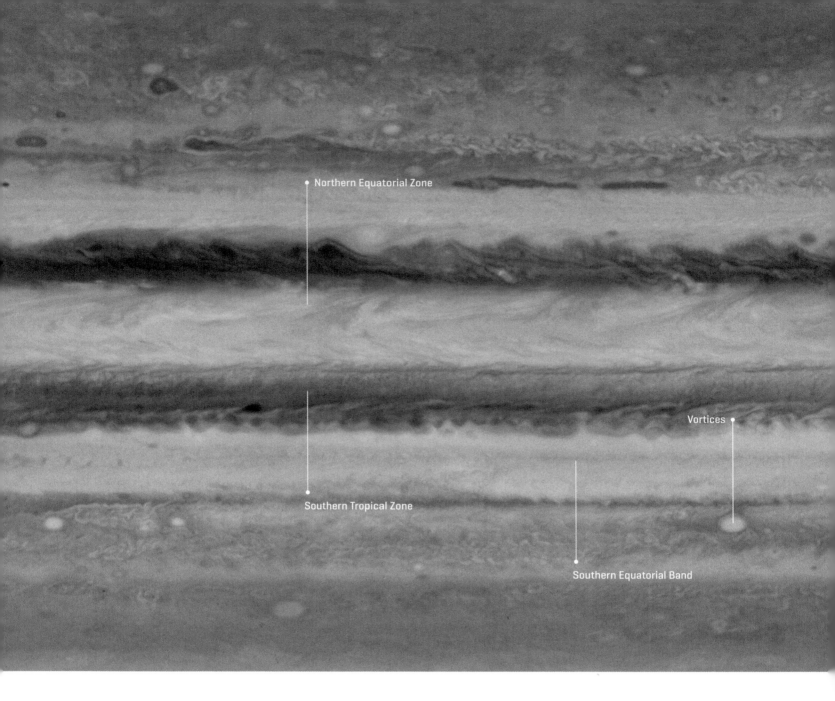

Northern Equatorial Zone

Vortices

Southern Tropical Zone

Southern Equatorial Band

MAPPING A GIANT

This cylindrical projection of Jupiter allows us to see the whole planet at a glance and highlights details like the Great Red Spot, its equatorial bands, and the planet's poles.

This projection captures Jupiter's entire surface in one rectangular image. While such a view is helpful when it comes to seeing central elements, such as the Great Red Spot, it presents a problem that mapmakers have long grappled with: the image distorts as it approaches the poles because it is impossible to perfectly represent a sphere's surface in a rectangular plane. To overcome this shortcoming, this map is complemented by two polar projections (on the opposite page), allowing us to appreciate both hemispheres with greater precision.

Phenomena That Escape the Projection
The changing nature of Jupiter's surface prevents static maps from showing some of its significant details. For example, the bands of ammonia clouds that define the planet's jets of air and large vortices (cyclones and anticyclones) are ever shifting, making it difficult to lock them down on the page.

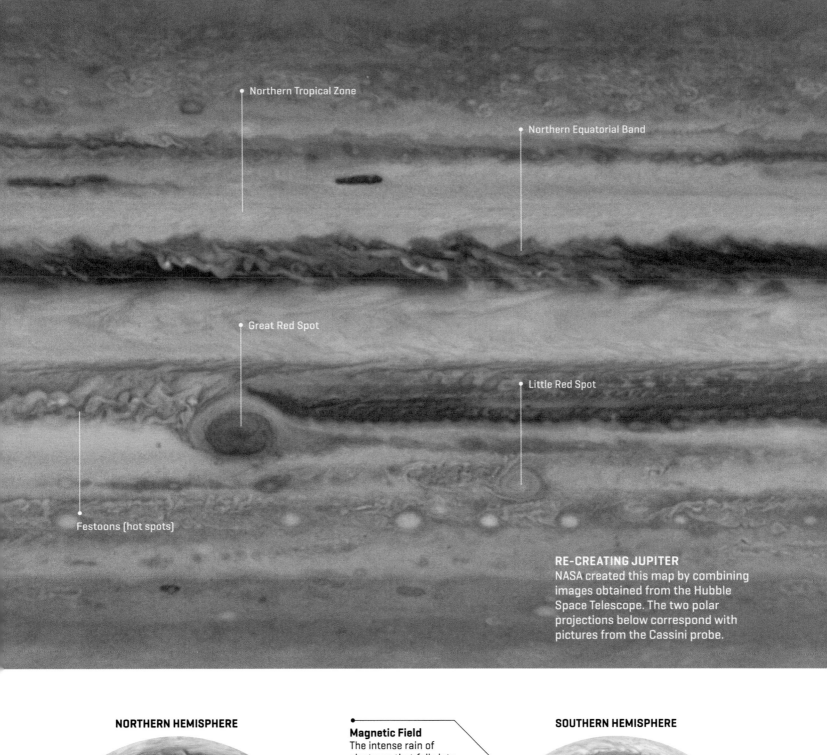

Northern Tropical Zone

Northern Equatorial Band

Great Red Spot

Little Red Spot

Festoons (hot spots)

RE-CREATING JUPITER
NASA created this map by combining images obtained from the Hubble Space Telescope. The two polar projections below correspond with pictures from the Cassini probe.

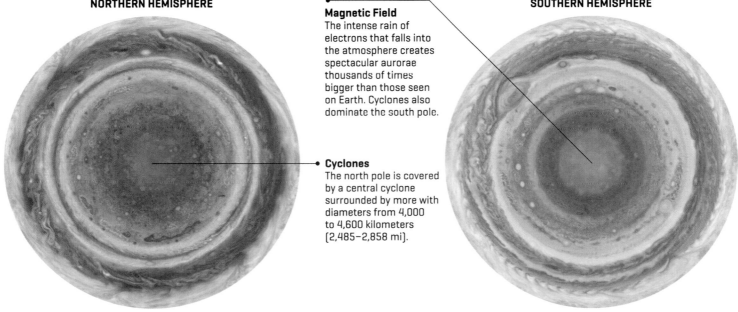

NORTHERN HEMISPHERE

SOUTHERN HEMISPHERE

Magnetic Field
The intense rain of electrons that falls into the atmosphere creates spectacular aurorae thousands of times bigger than those seen on Earth. Cyclones also dominate the south pole.

Cyclones
The north pole is covered by a central cyclone surrounded by more with diameters from 4,000 to 4,600 kilometers (2,485–2,858 mi).

SATURN, THE LORD OF THE RINGS

Saturn shares characteristics with the other gas planets: it's large, composed mostly of hydrogen and helium, and rotates quite quickly. Like the others, it also has a ring system—by far the most amazing in the solar system.

Saturn is the sixth closest planet to the sun, and far enough away from Earth that it is not very bright in our night sky. The rotational speed near the equator is 35,000 kilometers (21,748 mi) per hour, and winds in the upper atmosphere can reach up to 1,500 kilometers (932 mi) per hour. The resulting pressure on its surface layers is higher than the pressure found 1 kilometer (0.62 mi) deep in Earth's oceans. Only Jupiter is larger in both mass and size. It has the lowest density in the solar system, 0.7 g/cm³ (0.025 lb/in³): even less than that of water. The planet's magnetosphere, generated by its magnetic field and rotating core, is not as intense as Jupiter's. It is somewhat weaker than Earth's around the equator, but its influence extends some 20 times farther.

System of Rings and Moons

Saturn is well known for its ring system: seven million rings formed by fragments of ice and rock, and named with letters of the alphabet according to their order of discovery. Each of its 62 confirmed moons is a very unique world. The seven largest, which are ordered by size in the table opposite, always show Saturn the same face. The largest moon, Titan, has a landscape surprisingly similar to Earth's, complete with rivers, seas, and clouds, but with methane instead of water. Its remarkable atmosphere is four times denser than Earth's. Enceladus has an ocean beneath its icy crust, with hydrothermal vents on its south pole that could provide the conditions needed for life to exist.

Internal Structure

Atmosphere
Hydrogen and helium are its main components, with clouds formed through the condensation of ammonia, water, and other elements.

Lower Atmosphere
Hydrogen dominates at this level, in both gas and liquid form depending on its depth and corresponding pressure.

Core
Made of rocks and metals, it can reach temperatures near 12,000°C (21,632°F).

Layer Around the Core
The metallic hydrogen that makes up this layer is under extreme pressure, which creates the planet's intense magnetic field.

| Sun | Mercury | Venus | Earth | Mars | Jupiter | Saturn | Uranus | Neptune |

Rocky Planets

Gas Planets

A GIANT, CLOSE UP
This composite image taken by the Cassini probe in October 2004 is comprised of 102 photographs, offering us a clear view of the planet's northern hemisphere and its ring system.

ATMOSPHERE

- Hydrogen 96.3%
- Helium 3.3%
- Other components (primarily methane and ammonia) 0.4%

SATELLITE	DIAMETER (km)	MASS (kg)	ORBITAL PERIOD (days)	GRAVITY ON THE SURFACE (m/s²)	DISTANCE TO SATURN (km)
Titan	5,150	1.3×10^{23}	16	1.352	1.2 million
Rea	1,528	2.3×10^{21}	4.5	0.264	527,000
Iapetus	1,472	1.8×10^{21}	79	0.223	3.5 millon
Dione	1,124	1.1×10^{21}	2.7	0.232	377,400
Tethys	1,066	6.2×10^{20}	1.9	0.145	295,000
Enceladus	504	1.1×10^{20}	1.4	0.113	238,000
Mimas	396	3.7×10^{19}	0.9	0.064	186,000

SATURN AND EARTH

These two planets are separated by more than distance.
Many aspects set them apart, including their composition,
size, and rotational periods.

While Earth has a thin atmosphere, this gas giant's atmosphere makes up a significant chunk of the planet. On the other hand, Saturn's core, comparable in size to Earth's, is only a small part of its total volume. Atmospheric phenomena in Saturn's cloud layers are similar to those on Earth, including wind, lightning storms, and polar aurorae. Their surface gravity is also very similar, although their atmospheric pressure and temperatures are in very different ranges. Saturn has different seasons, although the climatic conditions change less because of Saturn's distance from the sun—summer is characterized by a slight increase in atmospheric temperature. Seasonal storms occur on Saturn once a year, which is equal to 30 Earth years. Its rotational speed is faster than Earth's: one day on Saturn is less than 11 hours long. At night, its rings reflect solar light that illuminate the dark side more brightly than our moon lights up Earth.

Rings and Moons
The Earth has only one moon, while Saturn has a complete system of rings and moons. If its main rings surrounded our planet, they would reach one-third of the way across the distance that separates us from our moon, while the total diameter of Saturn's system is 30 times the distance between the Earth and the moon. Some 200 objects have been observed around the planet, although we have only been able to confirm the orbits of 62 of them.

In Numbers	AREA	MASS	AVERAGE ORBITAL DISTANCE	ROTATION (DURATION OF A DAY)
Saturn	42.7 billion km² (16.5 billion mi²)	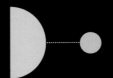 5.688 × 10²⁶ kg	 1,433 million km (890+ million mi)	10 h 39 min
Earth	510.1 million km² (197 million mi²)	5.97 × 10²⁴ kg	150 million km (58 million mi)	23 h 56 min

SATURN'S LARGEST MOON
Titan, the largest of Saturn's 62 known moons and the second largest in the solar system, is shown here orbiting the planet and its rings. Evidence suggests that Saturn may have more than 200 moons.

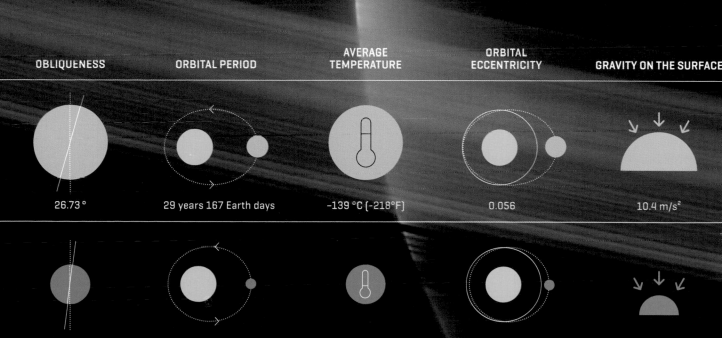

OBLIQUENESS	ORBITAL PERIOD	AVERAGE TEMPERATURE	ORBITAL ECCENTRICITY	GRAVITY ON THE SURFACE
26.73 °	29 years 167 Earth days	–139 °C (–218°F)	0.056	10.4 m/s²
23.44 °	365.25 Earth days	14 °C (57°F)	0.017	9.8 m/s²

THE RING STRUCTURE

Saturn's immense ring system has a peculiar, complex structure due to interactions with the planet and its moons. The moons settled along Saturn's equatorial plane, carving out gaps and keeping particles at the edges of the rings in place.

REVERBERATIONS IN THE RINGS

Orbital resonances impact the shape of Saturn's rings. They occur between bodies when they exert a regular, periodic influence on each other, with periods of rotation that are multiples of each other (meaning that they occasionally align). Whether or not moons have mutual gravitational attraction depends on their proximity, which varies with time. Particles in resonance with a satellite experience a force of maximum gravity at regular intervals that pushes and displaces them.

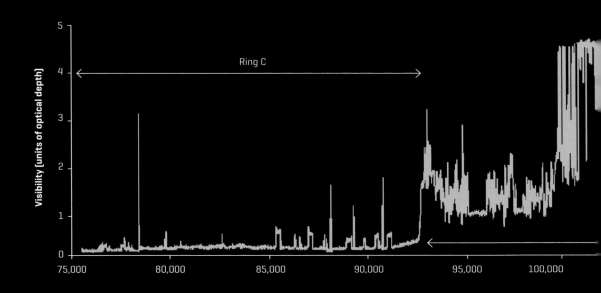

Ring C

Visibility (units of optical depth)

75,000   80,000   85,000   90,000   95,000   100,000

D   C

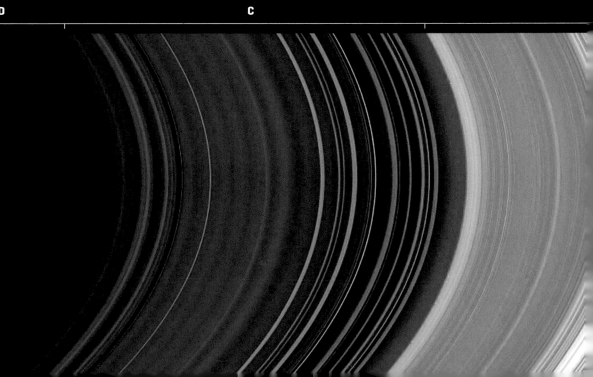

DIVIDING LINES
This illustration of Saturn's rings gives us a bird's-eye view of them, ordered according to their distance from Saturn's equator. The less bright zones appear to coincide with a lower particle density and create the divisions between rings.

Letters are used to label Saturn's rings in the order of their discovery, so A, B, and C correspond to the most visible. D is the closest to the planet, F and G are the thinnest, and E is the weakest and most extensive at 800,000 kilometers (497,097 mi). The particles that make up the rings are distributed in the planet's equatorial plane, but the density of each determines its respective brightness.

Radial Structures

Gaps between some of the rings come from the gravitational effect of satellites: the Encke division, a hole in ring A, and the Cassini division, between ring A (exterior) and B (interior). The latter, caused by the influence of the moon Mimas, has a density similar to that of ring D, but in contrast to A and B, it seems empty. Space probes that were able to get close to these formations revealed other surprises, such as radial structures in the B ring that reach lengths of 16,000 kilometers (9,942 mi) and, in a matter of hours, appear and disappear.

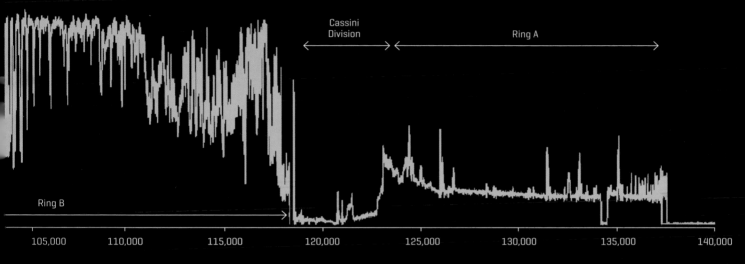

Cassini Division

Ring A

Ring B

105,000   110,000   115,000   120,000   125,000   130,000   135,000   140,000

Ring Plane Radius (km)

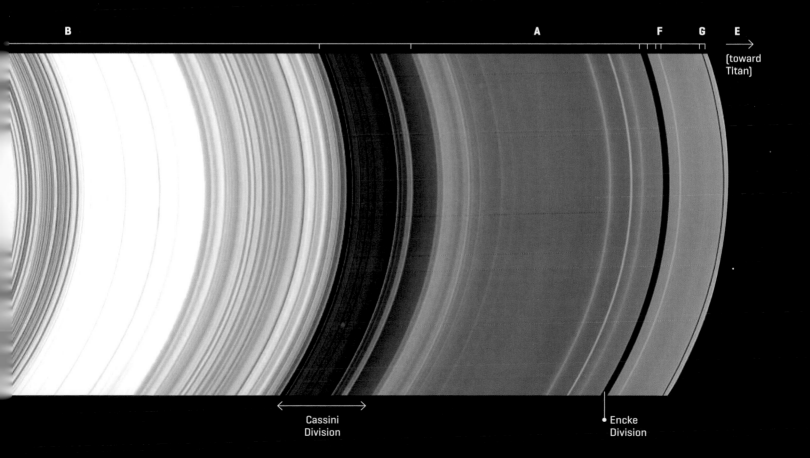

B   A   F   G   E
                                                             (toward Titan)

Cassini Division

Encke Division

SATURN IN THE SHADOWS
The Cassini probe captured this
mosaic image of Saturn blocking the
sun in 2006, showing an apparent
diameter at this distance (some 1,400
billion kilometers/869,917 million mi)
much less than its true diameter. The
photograph reveals Saturn's dim outer
rings, whose particles disperse light

LIFE IN THE OUTER SOLAR SYSTEM

Some gas giant moons have liquid water near their surface, and therefore may sustain some form of life. The most likely candidates are Enceladus and Europa, but Titan may be in the running as well.

Except for Io, most moons in the outer solar system are covered with thick crusts of ice. Scientists believe that below the ice of Europa (Jupiter's moon), Enceladus, and Titan (Saturn's moons) are large amounts of liquid water, which could allow for conditions similar to those of Lake Vostok in the Antarctic, where microbial life has been found.

Frigid Volcanoes
The key to the existence of life on these moons lies in cryovolcanism, or the ejection of water vapor from underground deposits, where heat created by gravitational forces causes the ice to melt. The structure of these ice volcanoes is almost identical to those of melted rock. Within them, liquid water acts like magma or lava, while ice acts like rock. In the case of Titan, Saturn's largest moon, water might be ejected along with methane.

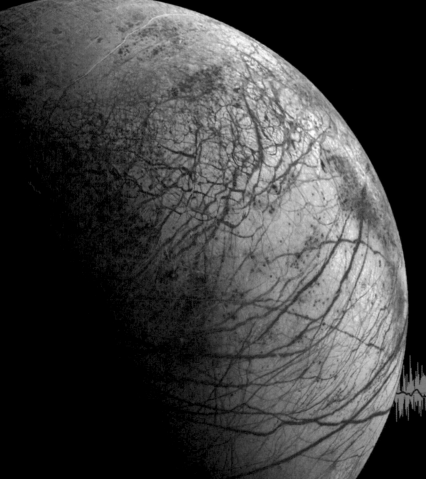

EXPLORING BENEATH EUROPA'S SURFACE

Recording Europa's seismic activity has allowed scientists to understand its structure and the possibility of finding life there.

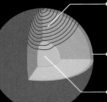

Quakes create specific signals, or "pings," which reveal the thickness of the crust.

The "pings" reach the ocean and measure its turbulence.

Compression waves reach the core and reveal its structure.

Oceanic whirlpools reveal its activity

Quakes in the crust reveal its thickness

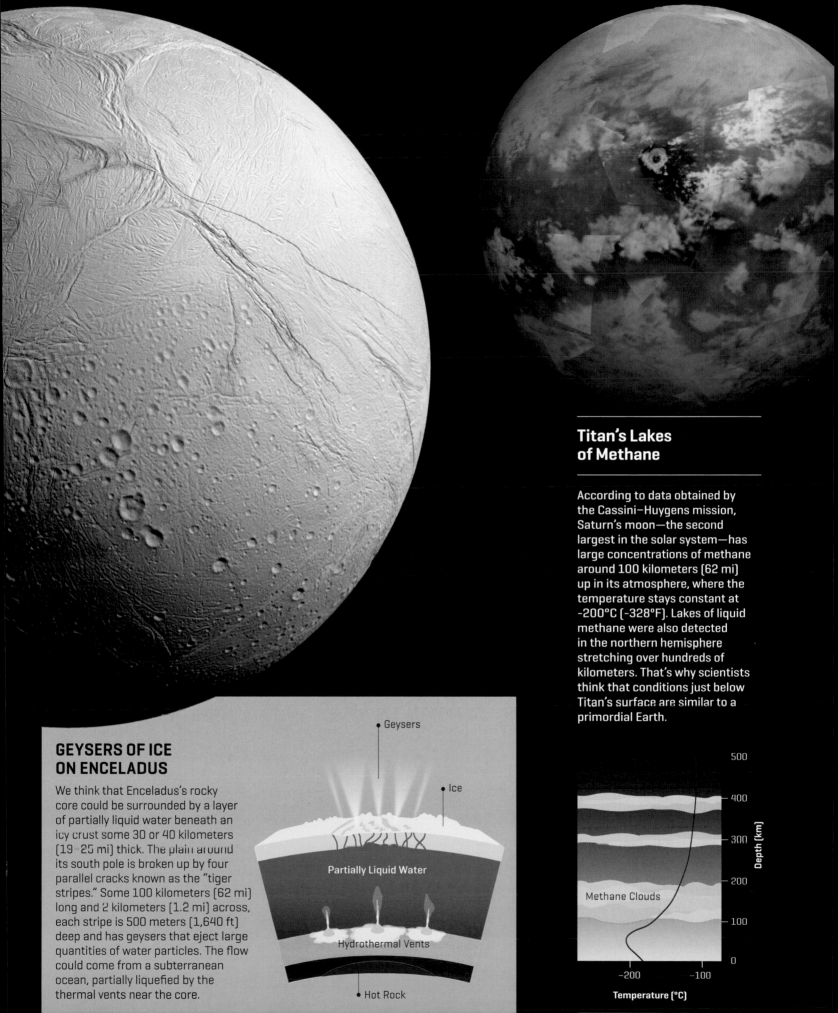

Titan's Lakes of Methane

According to data obtained by the Cassini–Huygens mission, Saturn's moon—the second largest in the solar system—has large concentrations of methane around 100 kilometers (62 mi) up in its atmosphere, where the temperature stays constant at -200°C (-328°F). Lakes of liquid methane were also detected in the northern hemisphere stretching over hundreds of kilometers. That's why scientists think that conditions just below Titan's surface are similar to a primordial Earth.

GEYSERS OF ICE ON ENCELADUS

We think that Enceladus's rocky core could be surrounded by a layer of partially liquid water beneath an icy crust some 30 or 40 kilometers (19–25 mi) thick. The plain around its south pole is broken up by four parallel cracks known as the "tiger stripes." Some 100 kilometers (62 mi) long and 2 kilometers (1.2 mi) across, each stripe is 500 meters (1,640 ft) deep and has geysers that eject large quantities of water particles. The flow could come from a subterranean ocean, partially liquefied by the thermal vents near the core.

Geysers

Ice

Partially Liquid Water

Hydrothermal Vents

Hot Rock

Methane Clouds

Depth (km)

500
400
300
200
100
0

Temperature (°C)

-200   -100

URANUS, AN ICY GIANT

The seventh planet from the sun and the third largest in the solar system is very cold and windy: a giant made of ice. Its most unique characteristic is the incline of its axis, which makes it appear to rotate on its side.

Uranus is similar to Neptune, made up mostly of ice and rock. But they are called ice, rather than gas, giants because they also contain water, ammonia, and methane ices. Sitting 2.872 billion kilometers (1.785 billion mi) from our star, Uranus has the fourth largest mass in the solar system and takes 42 years to make a trip around the sun. It has 13 rings and 27 known moons; only five, which appear in the table, were discovered before the advent of modern instruments. The angle between its axis of rotation and orbital axis around the sun is almost 98 degrees, so we see it tipped almost completely on its side—the planet's equator sits where the poles would normally be. Though it can actually be seen with the naked eye (though it is difficult to spot), it wasn't described as a planet (by William Herschel) until 1781. It had been observed before that but was mistaken for a star.

An Icy Atmosphere

The coldest of the planets in the solar system with a minimum temperature of −224°C (−371.2°F), Uranus does not have a true surface, as it consists primarily of swirling fluids that connect with the atmosphere and a complex structure of cloud layers.

MOON	DIAMETER (km/mi)	MASS (kg)	ORBITAL PERIOD (days)	GRAVITY ON THE SURFACE (m/s²)
Titania	1,577/980	3.527×10^{21}	8.71	0.38
Oberon	1,523/946	3.014×10^{21}	13.46	0.348
Umbriel	1,172/728	11.72×10^{20}	4.14	0.23
Ariel	1,158/720	1.35×10^{21}	2.52	0.27
Miranda	472/293	6.59×10^{19}	1.41	0.079

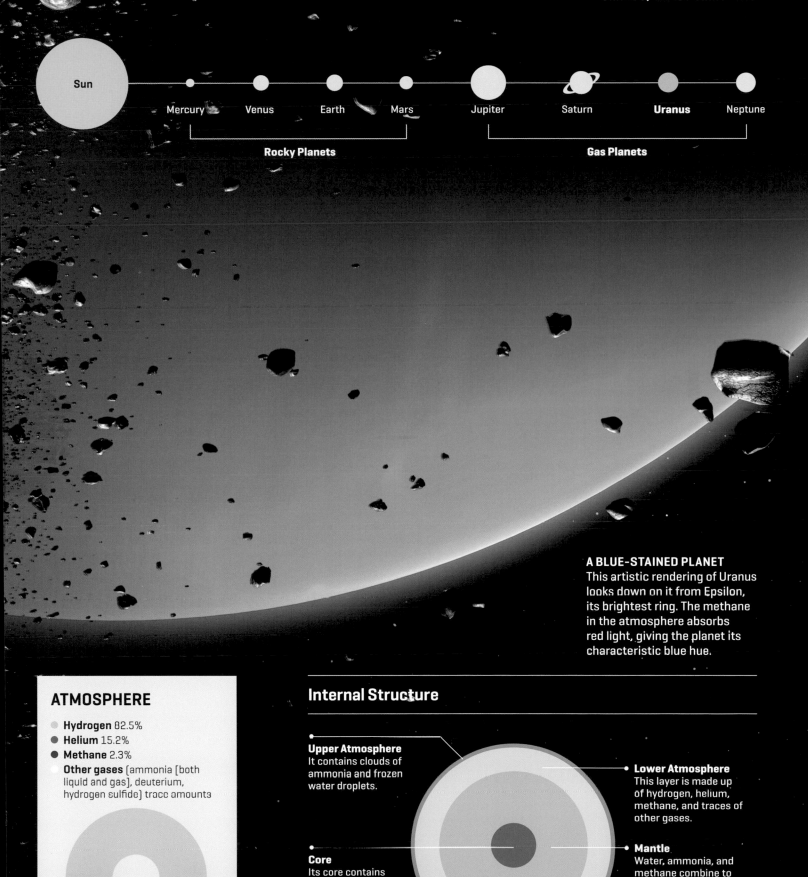

| Sun | Mercury | Venus | Earth | Mars | Jupiter | Saturn | **Uranus** | Neptune |

Rocky Planets

Gas Planets

A BLUE-STAINED PLANET
This artistic rendering of Uranus looks down on it from Epsilon, its brightest ring. The methane in the atmosphere absorbs red light, giving the planet its characteristic blue hue.

ATMOSPHERE

- **Hydrogen** 82.5%
- **Helium** 15.2%
- **Methane** 2.3%
- **Other gases** (ammonia [both liquid and gas], deuterium, hydrogen sulfide) trace amounts

Internal Structure

Upper Atmosphere
It contains clouds of ammonia and frozen water droplets.

Lower Atmosphere
This layer is made up of hydrogen, helium, methane, and traces of other gases.

Core
Its core contains a combination of silicates, solid iron, and nickel.

Mantle
Water, ammonia, and methane combine to produce a dense fluid whose streams create Uranus's magnetic field.

NEPTUNE, IN THE SYSTEM'S OUTER REACHES

Dark, cold, and plagued by supersonic winds, this densest of the giant planets is the farthest known planet from the sun and can't be seen without a telescope.

Neptune is far away from our star: more than 30 times farther than Earth. It is also four times larger and three times more massive. It has five rings and 14 moons, the largest of which are Triton, Proteus, and Nereid. Like Uranus, it does not have a true surface; its atmosphere goes down quite far, mixing with water and other fluids above a very heavy, solid core. It is the only planet in the solar system that cannot be observed with the naked eye and was only discovered in 1846 with some complicated calculus, then confirmed a year later with a telescope. Since then it has gone around the sun only once, as it needs 165 years to make the journey.

Hurricane-Force Winds

Neptune is the windiest planet in the solar system. These winds can be up to four times stronger than Jupiter's, reaching 2,000 kilometers (1,240+ mi) per hour—a speed even higher than Saturn's winds. They come from the heat made by Neptune's interior, which could contain leftover material from the planet's formation that is slowly leaking out. In 1989 the Voyager 2 probe photographed a storm there called the Great Dark Spot, which was large enough to contain our Earth. It has since disappeared, but there are new spots in other locations.

Internal Structure

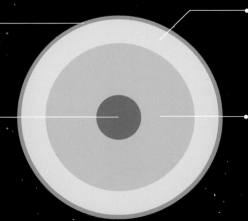

Upper Atmosphere
Clouds of ammonia and water condense in this layer.

Core
It is a solid ball of silicates, iron, and nickel, with pressure two times that of Earth's core.

Lower Atmosphere
This layer is made of hydrogen, helium, methane, and traces of other gases. Methane is part of what gives the planet its deep blue tint.

Mantle
The mantle is a mix of water, ammonia, and methane, and is where the majority of Neptune's mass can be found.

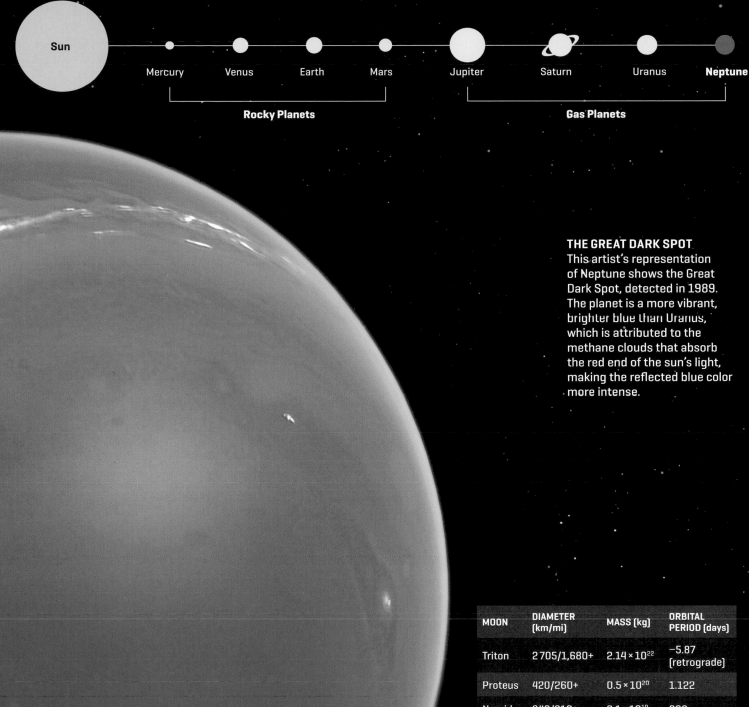

Sun · Mercury · Venus · Earth · Mars · Jupiter · Saturn · Uranus · **Neptune**

Rocky Planets · **Gas Planets**

THE GREAT DARK SPOT
This artist's representation of Neptune shows the Great Dark Spot, detected in 1989. The planet is a more vibrant, brighter blue than Uranus, which is attributed to the methane clouds that absorb the red end of the sun's light, making the reflected blue color more intense.

MOON	DIAMETER (km/mi)	MASS (kg)	ORBITAL PERIOD (days)
Triton	2705/1,680+	2.14×10^{22}	−5.87 (retrograde)
Proteus	420/260+	0.5×10^{20}	1.122
Nereid	340/210+	3.1×10^{19}	360

ATMOSPHERE

- Hydrogen 80%
- Helium 19%
- Methane 1%
- Other gases (deuterium, ethane) trace amounts

THE MINOR BODIES AND INTERPLANETARY SPACE

The Hale–Bopp comet approaches the sun in 1997. Discovered two years before, it was an extraordinarily bright comet that could be seen with the naked eye from both hemispheres for 18 months.

THE ASTEROID BELT

Most of the asteroids in the solar system are found in a ring
between Mars and Jupiter, ranging from simple specks of dust to
dwarf planets such as Ceres.

Originally, the asteroid belt was just
another swathe of planetesimals in
the protoplanetary disc. Under normal
circumstances, they would have collided
with each other and formed protoplanets,
but the gravitational influence of nearby
Jupiter quickened their orbits, making their
collisions more violent and ensuring that
they broke apart. Some of these objects
remained where they were, while others fell
toward the inner planets as massive meteor
showers. Most were left with eccentric
orbits, and the belt ended up with only 4
percent of the moon's mass.

Types of Asteroids According to Structure

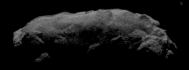

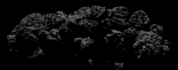

Compact Rocks
These are made up of a uniform
mass of rocks and densely
compacted dust.

Rock Clusters
Several small rocks are
combined, held weakly together
by compacted dust.

Fleets of Rocks
Groups of rocks are held
together by gravitational
equilibrium. Two-thirds of these

Solid Core
These asteroids are made up
of a central rocky fragment
surrounded by a halo of dust.

THE PASSAGE OF EUPHROSYNE
NASA's Wide-field Infrared Survey Explorer (WISE) space telescope took a series of images on May 17, 2010, of the fifth most massive asteroid in the belt: Euphrosyne. Overlapped, these images show its cosmic trajectory against a stellar background.

ASTEROIDS IN SYNC

Most asteroids are concentrated in the belt, but there are five points in Jupiter's orbit where the combined force that planet's gravitational pull (B) and the sun's (A) are in equilibrium, creating a point of balance that allows objects to hover there in a synchronous orbit. These are the so-called Lagrange points, which are numbered from L1 to L5 and have names like Hilda, Trojans, and Greeks.

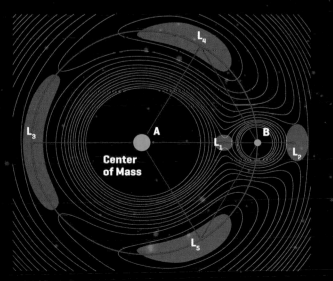

ROCKS IN SPACE
On either side of the asteroid belt are two larger bodies—Vesta to the left and Ceres to the right—that account for more than 40 percent of the belt's total mass. This illustration shows only a fraction of the asteroid ring's density.

THE KUIPER BELT AND THE OORT CLOUD

When Pluto was first discovered, there was speculation about the existence of two regions beyond its orbit from which comets might come: one similar to the asteroid belt, and another, gigantic cloud that would envelop the entire solar system.

This hypothesis sprang from the need to explain where comets come from. The Kuiper belt was thought to be a hatchery for the closest comets, and its existence was confirmed in 1992. It extends from Neptune's orbit 30 AU from the sun to about 50 AU and is populated by large bodies that include Pluto, Haumea, Makemake, and other small, irregular bodies. Its total estimated mass is some 20 to 200 times that of the main asteroid belt.

The existence of the Oort cloud is still speculative, as it has yet to be observed, but we think it may be where the most distant comets come from. Astronomers estimate that this theoretically spherical region could have between one billion and 100 billion bodies, with a total mass some five times greater than Earth's and an outer limit more than 1.5 light-years from the sun.

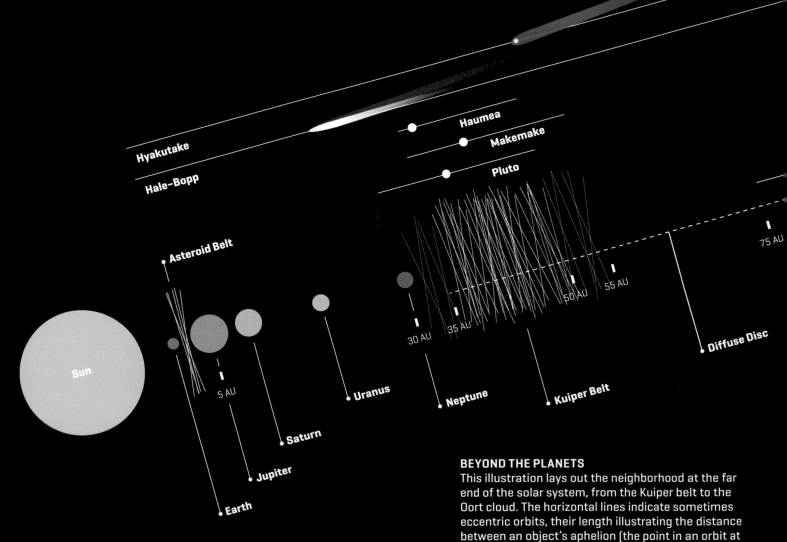

BEYOND THE PLANETS
This illustration lays out the neighborhood at the far end of the solar system, from the Kuiper belt to the Oort cloud. The horizontal lines indicate sometimes eccentric orbits, their length illustrating the distance between an object's aphelion (the point in an orbit at which an object is farthest from the sun) and perihelion (the point at which an object is closest to the sun).

CLOUD VISITORS

The closest comets come from a disperse disc-shaped structure that extends out from the sun some 460 AU. Scientists believe that the farthest of them could have come from the Oort cloud, most likely because of the galaxy's gravitational pull.

Oort Cloud

100,000 AU

Sedna

900 AU

460 AU

A GARGANTUAN SPHERE

The Oort cloud is theorized to be a disc that becomes increasingly spherical the farther it gets from the sun—a bubble that encompasses our solar system. The most remote part of this giant shell is some 100,000 AU away, which means it cannot be observed directly.

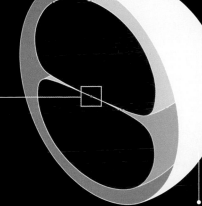

Outer Solar System

Kuiper Belt

Oort Cloud

①

②

DWARF PLANETS

These planetary bodies are of considerable size but lack orbital dominance, disqualifying them from joining the ranks of the eight larger planets.

Pluto and Charon

1 Identified through photos in 1930, Pluto was considered the solar system's ninth planet until 2006, when it was downgraded to a dwarf planet. It is 39 AU from the sun and rotates along with its large moon, Charon. Its mass is equivalent to 5 percent of Mercury's, and it is thought to have a dense, rocky core surrounded by a mantle of water. Its plains and atmosphere include nitrogen, methane, and carbon.

Ceres

2 Discovered in 1801 by Giuseppe Piazza, Ceres lost its status as a planet a few decades later and was reclassified as a dwarf in 2006. It is the largest astronomical object in the main asteroid belt, measuring almost 950 kilometers (590 miles) across and comprising 30 percent of the belt's estimated total mass—though some thousand times less than the total mass of Earth. Its approximate distance from the sun is 2.77 AU.

Sedna

3 Sedna, officially called 2003 VB12, takes its name from the Inuit god of the sea. It is registered as a plutoid, or ice dwarf: one of the smaller bodies in an orbital resonance with Neptune. Little is known about it except that its orbit is eccentric, and it does not come closer than 80 AU to the sun. We can tell it's a dwarf by its diameter, which is larger than Ceres's, and it is the farthest dwarf that we know of.

③

④

Eris

4 Discovered in 2005 and originally put forward as the 10th planet of the solar system, Eris is now the largest of the dwarfs and the largest body yet explored by a space probe. Its mass is more than 27 percent of Pluto's. Its moon, Dysnomia, is some 96.3 AU from the sun.

Some Trans-Neptunian Bodies

Pluto   Charon   Eris   Haumea   2007 OR10   Makemake   50000 Quaoar   Sedna

Pluto is the largest of the trans-Neptunian objects, with a diameter of 2,380 kilometers (1,479 mi). Next, from largest to smallest, are Eris (2,326 km/1,445 mi), Haumea (1,632 km/1,014 mi), 2007 OR10 (1,535 km/954 mi), Makemake (1,430 km/889 mi), 50000 Quaoar (1,110 km/690 mi), and Sedna (995 km/618 mi). All of them are likely dwarf planets.

COMETS

These bodies are some of the solar system's most distant. They are made mostly of rock and ice that melts as the comets near the sun, which is what gives them their spectacular and characteristic "tails."

Comets are classified according to their orbital period. Those with orbital periods less than 200 years are referred to as short-period, or periodic; those with orbital periods more than 200 years are called long-period, or nonperiodic. Some nonperiodic comets may be on orbits that take thousands, or even millions, of years to pass into the solar system. Short-period comets whose orbits are controlled by Jupiter's gravitational pull are referred to as comets from the Jupiter family. Other short-period comets, called great comets, are exceptionally bright and visible to the naked eye, with an orbital period of approximately 10 years. They have a large, active core (called their nucleus), a perihelion close to the sun, and a trajectory similar to Earth's. In 1705 English astronomer Edmond Halley was the first to calculate a comet's orbital period: 76 years. That comet—Halley's comet—was named after him.

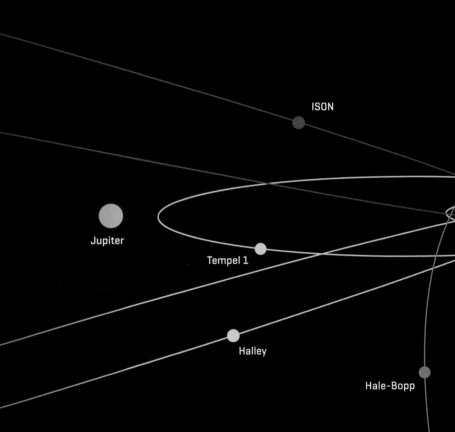

ISON

Jupiter

Tempel 1

Halley

Hale-Bopp

LANDING ON A COMET

The main probe of the European Space Agency's Rosetta mission, launched in 2004, orbited the sun four times and reached the 67P/Churyumov–Gerasimenko comet in 2014. Its module, Philae, landed on the comet successfully, but limited solar power affected its battery and reduced data collection. As the comet started to travel farther away, mission specialists opted for a controlled crash onto the comet's surface in 2016.

FROM HEAD TO TAIL

As a comet approaches the sun, its icy nucleus heats up and begins to melt, creating a kind of atmosphere. The solar wind pushes the ionized gas, creating its characteristic tail. The tail does not always trail behind the comet: because the solar wind dictates its direction, it always points away from the sun.

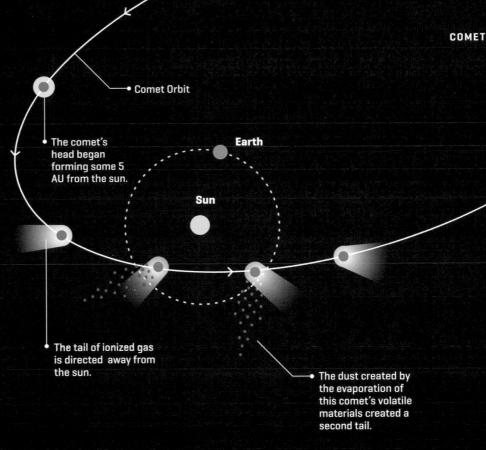

Comet Orbit

The comet's head began forming some 5 AU from the sun.

Earth

Sun

The tail of ionized gas is directed away from the sun.

The dust created by the evaporation of this comet's volatile materials created a second tail.

Sun

Earth

Churyumov-Gerasimenko

Hyakutake

RECENT COMETS

At the end of the 20th century, Hyakutake (seen in 1996) and Hale–Bopp (seen in 1997) put an end to a long comet dry spell. In 2005 NASA's Deep Impact space probe released an impactor into Tempel 1, a Jupiter-family comet that orbits the sun every 5.56 years, becoming the first spacecraft to eject material from a comet's surface.

- Nonperiodic Comets
- Comets from the Jupiter Family
- Long-period comets
- Short-period comets

THE HELIOSPHERE

The solar wind's range is like a sphere encompassing our planetary system, with a radius larger than 100 AU. Beyond it lies interstellar space.

The sun constantly emits a stream of protons and other charged particles in the form of solar wind. This wind extends from the sun in all directions and, apart from when it encounters interstellar streams of hydrogen and helium, forms a giant sphere. If we imagine ourselves as a particle in this wind, at some 80 to 100 AU from the sun we would experience a sudden deceleration to subsonic speeds (called the termination shock). We would end up traveling into the heliosheath, where our velocity would fall to zero. If we're in the "frontal" part of the heliosheath, depending on the sun's direction of rotation around the galactic center, the temperature would increase. Some 121 AU from the sun we would cross the heliopause, or the outer limit of the heliosphere, entering the interstellar medium and evaporating.

Valuable Estimations

We know these things about the solar wind thanks to the Voyager 1 and 2 probes, which went through the termination shock in December 2004 and 2007, respectively. Voyager 1 was the first human creation to enter interstellar space; it crossed the heliopause on August 25, 2012.

Interstellar Magnetic Field

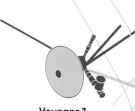

Voyager 1

121 AU

Heliosheath

Interstellar Stream

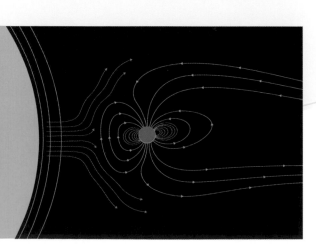

THE MAGNETOSPHERE, A SHIELD AGAINST SOLAR WIND

The solar wind is made of charged particles that interact with other magnetic fields. The Earth's magnetic field (or magnetosphere) diverts most of it, but some slips through at the magnetic poles and creates aurorae in the atmosphere. The most violent solar wind disturbances come from solar storms.

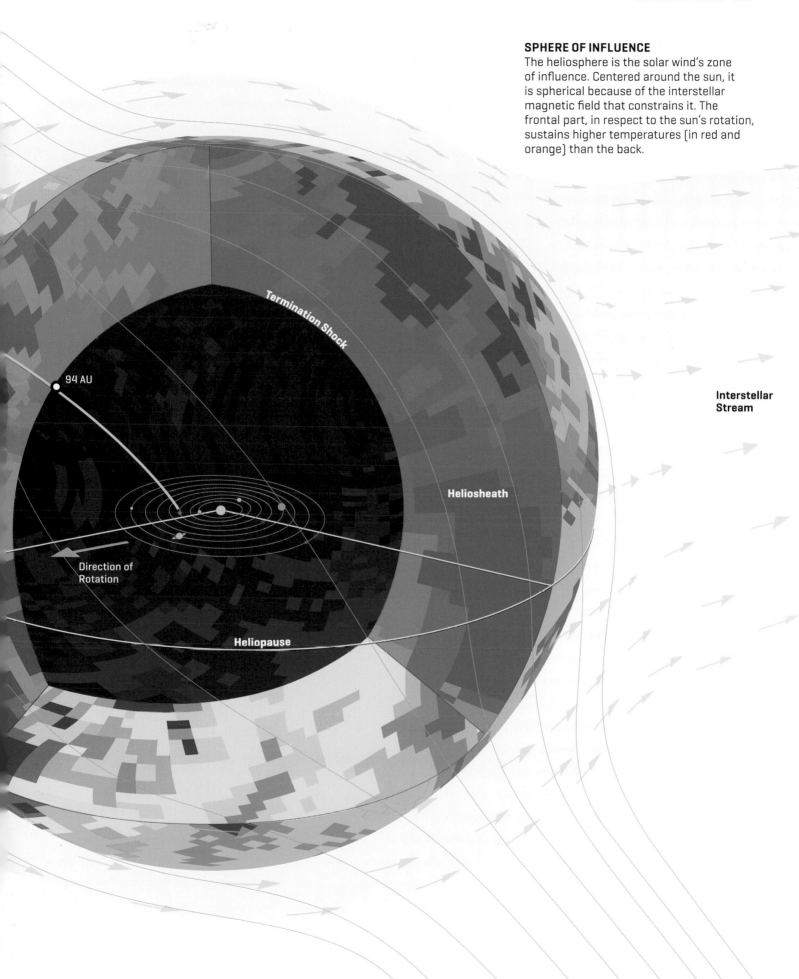

SPHERE OF INFLUENCE
The heliosphere is the solar wind's zone of influence. Centered around the sun, it is spherical because of the interstellar magnetic field that constrains it. The frontal part, in respect to the sun's rotation, sustains higher temperatures (in red and orange) than the back.

Termination Shock

94 AU

Heliosheath

Interstellar Stream

Direction of Rotation

Heliopause

OTHER

EARTHS,
OTHER SUNS

PLANETARY SYSTEMS

Until a few decades ago, the only planetary system we knew anything about was the solar system. Today we know about thousands of such systems, although they are only a fraction of what we believe exists out there in the universe. Both the stars and the planets that orbit them come from molecular clouds, just as the planets in our system did.

There are many billions of exoplanets in the Milky Way, and it's been calculated that some 10 billion of them should have similar characteristics to Earth and orbit stars similar to our sun—making them capable of supporting some kind of life. But given how vast interstellar distances are, it would be very difficult to reach or interact with them. Even if some kind of civilization does exist on one of the closest of these planets, it would take us decades to send messages back and forth.

Origin of the Molecular Clouds

The Milky Way's stars, and the planets that orbit them, formed from the same molecular clouds. Expansion waves caused by supernovae may be what triggered them to form celestial bodies. When the waves reach the cloud, it begins to flatten out and increase its rotational speed to conserve angular momentum.

Why Is It Disc-shaped?

The molecular clouds that make up the seeds of protoplanetary discs spin around an axis, while dust particles go around the cloud's center of mass. With such an enormous number of moving particles, collisions are inevitable. These collisions make the clouds slow down, decreasing their angles with respect to the orbital plane. The result is a concentration of particles that form a disc.

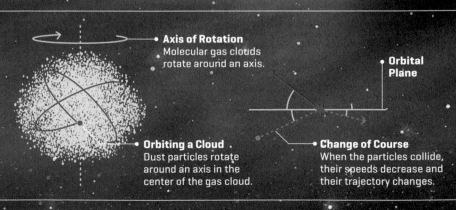

Axis of Rotation
Molecular gas clouds rotate around an axis.

Orbital Plane

Orbiting a Cloud
Dust particles rotate around an axis in the center of the gas cloud.

Change of Course
When the particles collide, their speeds decrease and their trajectory changes.

PROTOPLANETARY DISCS

Accretion discs rotate around a young star and can form planets. This image depicts protoplanetary disc Beta Pictoris, a star located some 60 light-years away from Earth. Its planetary system is in a phase similar to the first stages of the solar system, with signs that a giant planet is orbiting the star and that rocky planets and comets are being formed in the disc.

THE TRAPPIST-1 SYSTEM

In 2017, this system, located some 40 light-years from Earth, was revealed. TRAPPIST-1 is a red dwarf with at least seven planets orbiting it at a distance smaller than that between Mercury and the sun. However, due to the star's diminutive size, three of them are in the system's habitable zone, and their sizes do not differ much from Earth's.

OTHER WORLDS
This artist's conception shows a planet outside our system, as seen from its moon, and a binary star system lighting it up. Although the first exoplanet wasn't discovered until the end of the 20th century, advances in detection technology have allowed us to find thousands of bodies orbiting around their stars.

CIRCUMSTELLAR HABITABLE ZONE

No one knows exactly what life might look like on other planets, but one thing is certain: it has to obey the laws of chemistry.

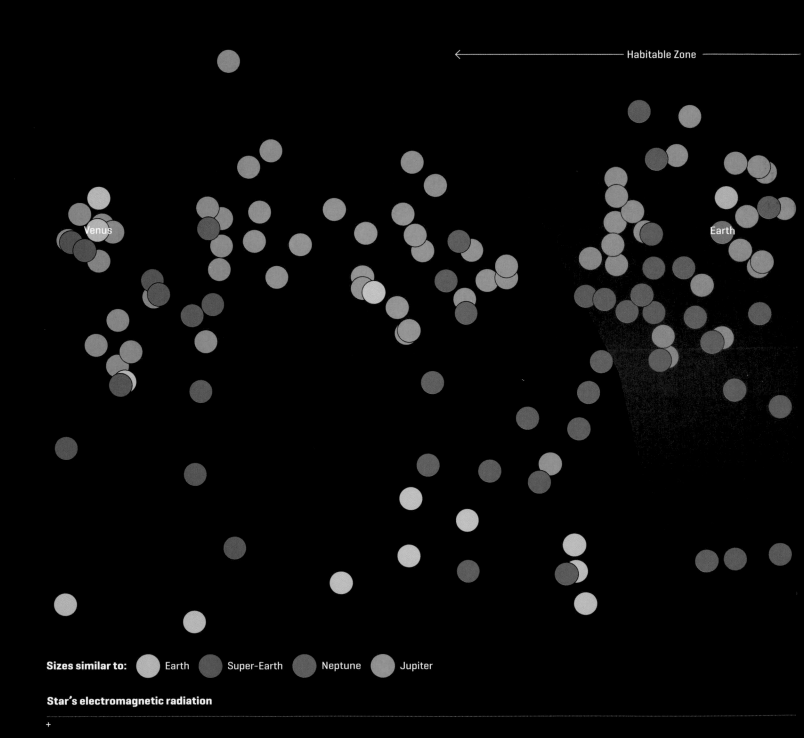

Sizes similar to: Earth Super-Earth Neptune Jupiter

Star's electromagnetic radiation

Molecules that may turn into life have an easier time reacting with each other when dissolved in a liquid. Water is ideal for this, as it is capable of dissolving a large number of different compounds. That's why astronomers think we're likely to find life in the so-called habitable zone of a star, or the area where water can exist as a liquid on a planet's surface: between 0°C and 100°C (32–212°F).

Neither Too Hot Nor Too Cold

This temperature range is essential for life because, although some substances become liquids at higher and lower temperatures, many organic molecules lose their stability in hotter environments. Consider, too, that chemical reactions tend to slow down under very cold conditions, making life unlikely. The graph below shows a large number of exoplanets, all classified by their respective locations in the habitable zone of their stars. A "super-earth" is an exoplanet that is much larger than Earth, but smaller than Neptune..

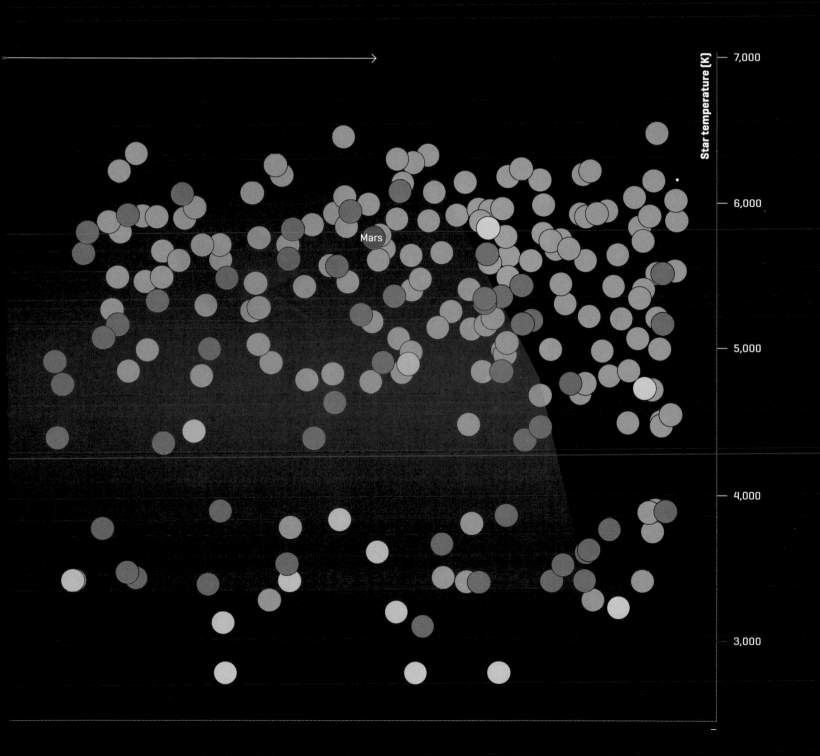

Star temperature [K]

7,000

6,000

Mars

5,000

4,000

3,000

GALACTIC HABITABLE ZONE

A planet's potential for hosting life is dependent not only on its position within its star system but also on its position in the galaxy.

According to recent research, the Milky Way has between 200 billion and 400 billion stars. Although not all stars have orbital planets, even low estimates suggest that there are at least 100 million in our galaxy that do. But not all are potentially habitable: that depends on whether they're in the habitable zone of their star and where the planetary system is within the galaxy. Scientists believe that systems close to the galactic bulge or near globular clusters have little possibility of developing life on some of the planets because they are bombarded with intense radiation, and their nearness to potential supernovae would wipe out all life.

Not Too Early or Close to the Center

The graph below shows the conditions of the Milky Way from its birth up to the present, as well as its distance from the galaxy's core. Very early and very close to the center, too few metals and too many supernovae prohibited planet formation. On the other hand, if it was too far from the center of the galaxy, there would not have been enough heavy metals to create rocky planets. The optimal habitable zone could not have existed early on in the Milky Way's existence or too close to its center.

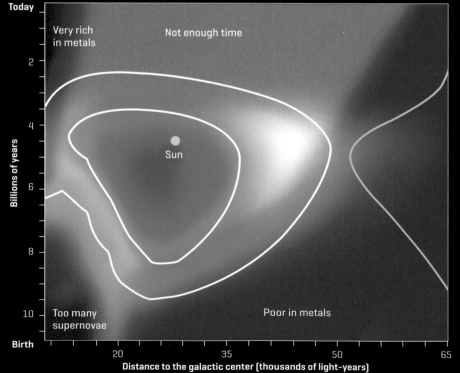

The Optimal Zone
In this graph of the Milky Way's position over time, the green zone represents the optimal habitable conditions, while the other colors show which parts are unlikely to support life and why.

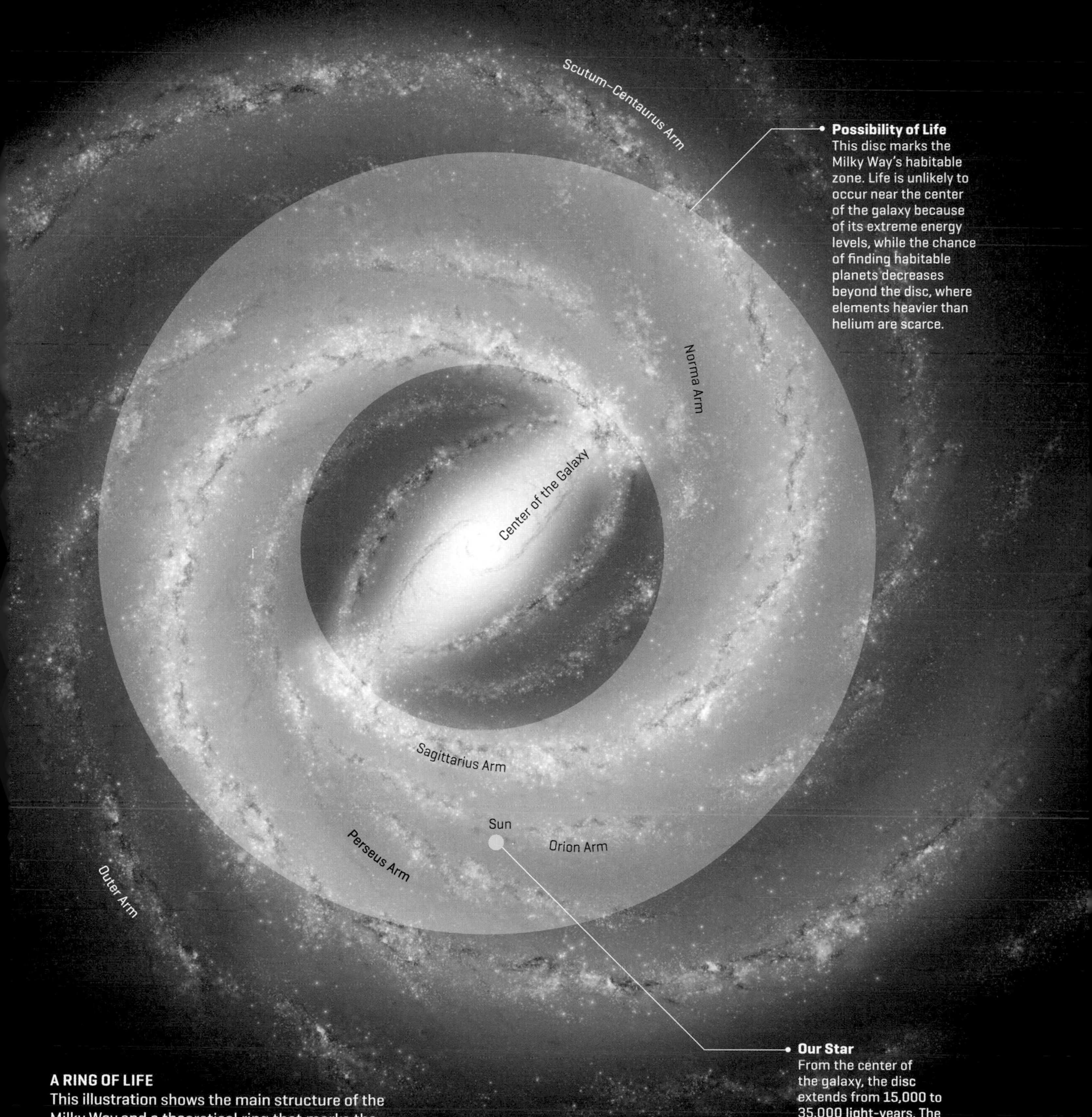

Scutum–Centaurus Arm

Norma Arm

Center of the Galaxy

Sagittarius Arm

Sun

Orion Arm

Perseus Arm

Outer Arm

Possibility of Life
This disc marks the Milky Way's habitable zone. Life is unlikely to occur near the center of the galaxy because of its extreme energy levels, while the chance of finding habitable planets decreases beyond the disc, where elements heavier than helium are scarce.

Our Star
From the center of the galaxy, the disc extends from 15,000 to 35,000 light-years. The sun is some 25,700 light-years from the center of the galaxy.

A RING OF LIFE
This illustration shows the main structure of the Milky Way and a theoretical ring that marks the habitable zone. Our solar system is in a privileged position, practically in the center of the zone.

INDIRECT METHODS OF DETECTION

It's difficult to directly observe planets outside our solar system, so we have to rely on indirect methods of detection.

Discovering planets outside the solar system, called exoplanets, is an extremely complicated task. Their inherently faint light is almost impossible to see through the enormous distances between them and us. Their apparent size makes their silhouettes too small to see through a telescope, and the light from the stars is so intense that it covers all traces of the planets orbiting them. Still, some large, young exoplanets have an intrinsic brightness that makes them easier to find.

Difficulties of Direct Observation
Exoplanets appear as small points of light (visible or infrared) around the stars. As a result, it is usually impossible to observe their reflected light or infrared radiation directly. That requires indirect methods of detection, as shown in the illustrations below.

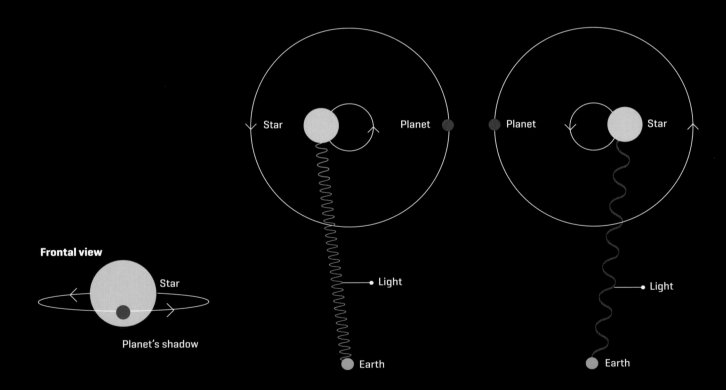

Frontal view

Star

Planet's shadow

Light

Earth

Light

Earth

1. Transit
We observe the star's brightness for any dimming, which could be a sign of an orbiting planet passing partially in front of it.

2. Radial Speed
Light becomes bluer when the source emitting it comes closer to us, and redder when it moves away. If a star is surrounded by planets, we can see their movement around the system's center of gravity because of these changes in light.

LIGHT'S INFLUENCE
An exoplanet's silhouette
cannot be observed directly,
but we can gain information
about its size and
atmosphere from the light
its stars transmit through its
atmosphere, as pictured in
this illustration.

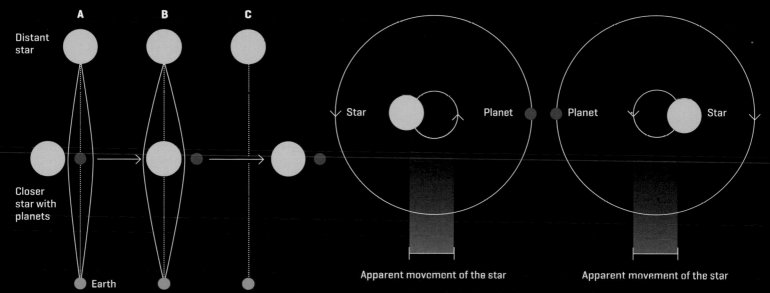

3. Gravitational Microlensing
When one star passes in front of another (from
Earth's point of view), the gravity of the closer
star curves the light of the farther star and
increases its brightness. If it has a planet, this
can cause an intense flash.

4. Astrometry
We can detect the presence of a planet
whose planetary system's center of gravity
is sufficiently far from the star by precisely
measuring small changes in the star's positions.

STARS WITH EXOPLANETS

Because detecting exoplanets is a complicated task, the field suffers from the occasional false positive. Even so, thousands of planetary systems have been discovered in the last few decades.

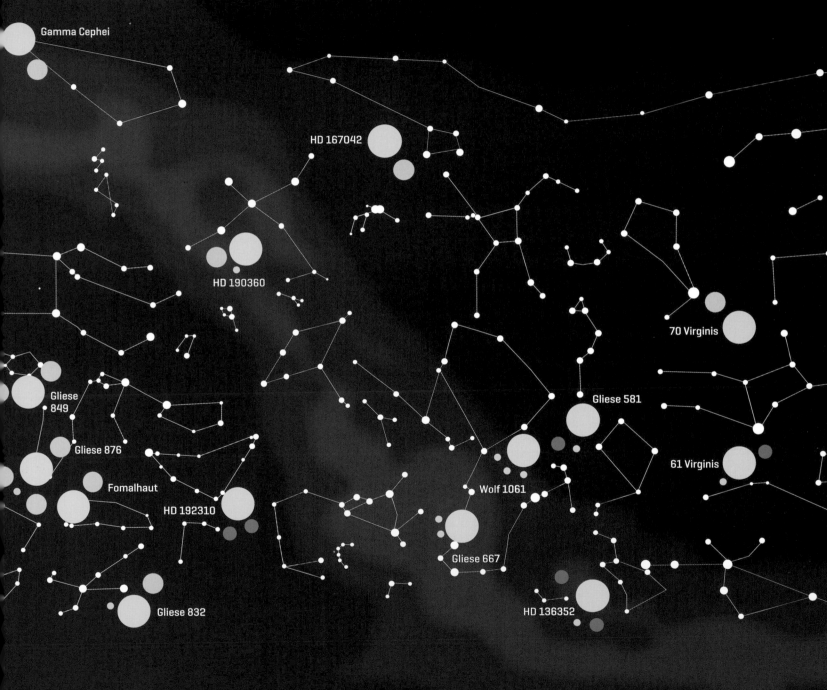

Gamma Cephei

HD 167042

HD 190360

70 Virginis

Gliese 849

Gliese 581

Gliese 876

61 Virginis

Fomalhaut

Wolf 1061

HD 192310

Gliese 667

Gliese 832

HD 136352

Proxima Centauri

- Earth-type
- Neptune-type
- Jupiter-type

Even after its discovery, an exoplanet does not become part of the catalog of known worlds until further observations can confirm its existence. Though thousands of planets outside our solar system have been found in recent decades, the signals that our instruments detect often turn out to be false positives. A star's temporary reduction can be due to a simple change on its surface, for example, or perhaps an eclipse of a binary star system. Gliese 667C is a perfect example: the star was believed to have seven orbiting planets, but later observations showed that five of them were actually noise recorded during the measurements.

Nearby Stars

Even so, nearby stars with exoplanets orbiting them are still being discovered. Some of them, like 55 Cancri and 47 Ursae Majoris, are visible to the naked eye.

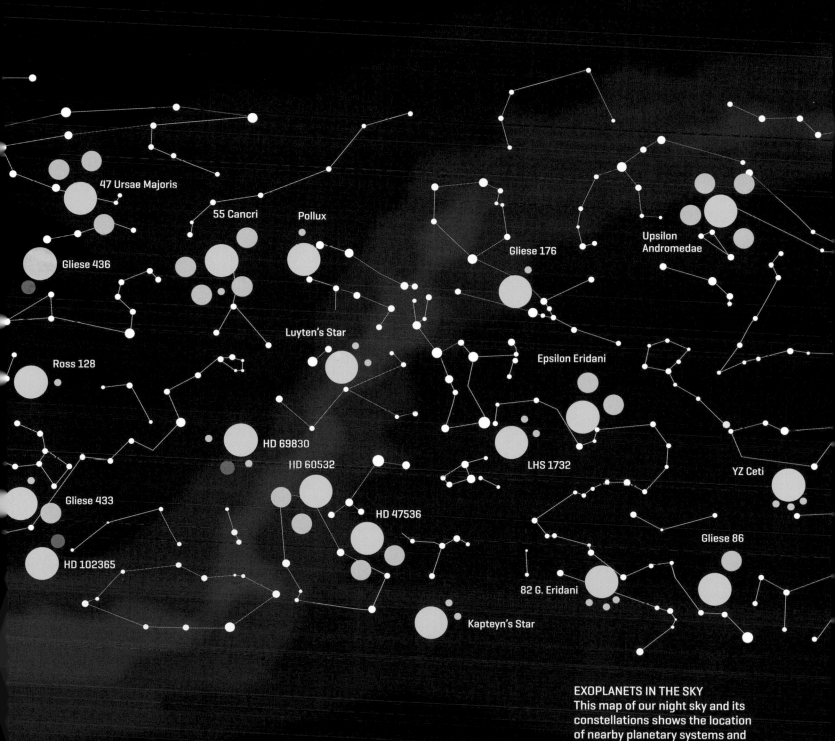

EXOPLANETS IN THE SKY
This map of our night sky and its constellations shows the location of nearby planetary systems and their confirmed exoplanets.

TYPES OF EXOPLANETS

In general terms, there are five types of exoplanets, categorized by how their mass compares with known bodies in the solar system.

It's hard to find out much more about exoplanets than their mass, size, and orbit, but this information can tell us a lot about them. Astronomers can take the mass of an exoplanet and have a good idea of whether it is a small, rocky world like Earth or a gas giant like Jupiter.

Classified by Masses

Although the edges of each category are fuzzy, exoplanets can be grouped into five types according to their mass: the largest (similar to Jupiter, or even heavier); those similar in size to Neptune; "super-Earths," which are slightly larger than Earth; planets similar to Earth; and smaller terrestrial bodies called "sub-Earths."

MASS AND COMPOSITION
The nature of exoplanets can vary widely according to their composition. Astronomers take care when classifying a planet as one type or another, especially with objects whose masses put them on the line between super-Earths and Neptune-type planets.

Jupiter-type Planets
These gas giants are between 60 and 4,000 times more massive than Earth, with a diameter between five and seven times greater. They are easy to identify due to their mass and size.

Neptune-type Planets
These objects are more massive than Earth but less so than Jupiter, with a mass less than the lower limit of a gas giant, but heavier than 10 Earth masses.

Super-Earths
These planets have a mass less than 10 Earth masses, but they are larger than Earth.

Earths
Also known as exo-Earths, these exoplanets have a mass similar to our home planet.

Sub-Earth
These planets have a mass less than that of Earth or Venus. Mercury and Mars fit into this classification.

POTENTIALLY HABITABLE PLANETS

Evaluating whether or not exoplanets have the necessary conditions for life requires very precise data about their composition, their orbit, and especially their atmosphere.

Among the thousands of exoplanets we've discovered, some are similar to Earth and may be habitable. Given the few details we have about their compositions and atmospheres, we assess their habitability according to whether they are rocky planets and are far enough from their star that liquid water can exist on their surfaces. However, these two factors do not guarantee that they have favorable conditions for life. For example, Venus is in the habitable zone but was rendered unsuited for life due to its composition; a planet could be in a cold enough region, but reach high temperatures because of how well its atmosphere retains heat. Scientists believe that a number of exoplanets could be habitable, but that list might shrink once we measure their atmospheres.

Habitable Planetary Systems

Of the planetary systems that could have the conditions for life, one shows particular promise: TRAPPIST-1. This system has seven rocky planets, three of which are in the habitable zone of its star. Until recently, the Gliese 667C system was also thought to have seven planets in its habitable zone, but further exploration reduced that number to two. Kepler-62 also has a planet in its stellar habitable zone.

Kepler-62

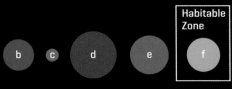

Habitable Zone

TRAPPIST-1

Habitable Zone

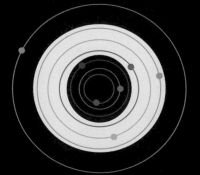

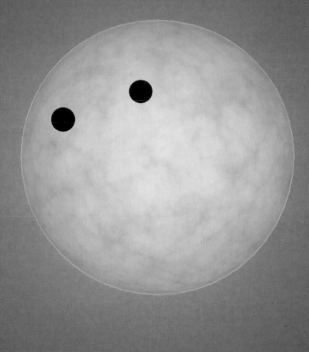

TWILIGHT ON EARTH AND EXOPLANETS

This artist's rendering compares a sunset on Earth (1) with those you might see on exoplanets thought to have a high probability of life. Sunsets seem to be redder on the largest planet, Gliese 667Cc (2), and the most distant, Gliese 581d (5), because they orbit red dwarfs. Kepler-22b (3) has a sunset like Earth's because it orbits a star like our sun, but HD 85512b (4) has the brightest of all since its star, a K dwarf, is hotter.

CLOSE PROXIMITY
The planets are close enough to each other in the TRAPPIST-1 system that, if you were to look up from the surface of one of them, the others would appear twice as big as our sun as we see it from Earth.

1   2   3   4   5

THE SEARCH FOR LIFE

Life tends to leave traces behind in its surroundings, which makes it possible to look for traces of it in a planet's atmosphere.

Discovering living organisms or fossils on another planet would finally answer a pressing scientific question: does extraterrestrial life exist? But until we can visit other planetary systems, scientists have to use indirect ways to seek out life elsewhere.

Chemical Imbalances
Sooner or later, an inert planet will reach a chemical imbalance in its atmosphere, and very reactive elements like oxygen will combine with other materials until there is absolutely no sign of them. Thus large quantities of oxygen on an exoplanet could be the product of simple chemical reactions, not a promising sign of life.

A SUPER-EARTH
Exoplanet Gliese 667Cc, a super-Earth, is found in the habitable zone of its solar system. It orbits a red dwarf star, Gliese 667C, which is part of a triple star system nearly 24 light-years away from Earth in the Scorpius constellation.

Signs of Life in the Electromagnetic Spectrum

We can deduce the composition of an exoplanet's atmosphere by analyzing the electromagnetic spectrum of its light. Oxygen and methane are particularly valuable as a possible sign of life, along with other compounds such as chloromethane and dimethyl sulfide. The following illustrations show the emission spectra of Mars and Venus compared to that of Earth, which has both oxygen and water molecules.

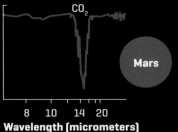

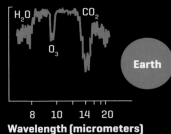

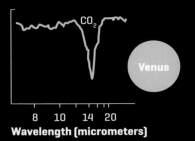

THE EXTREMES OF LIFE ON EARTH

Earth's organisms show how life can adapt to a wide range of conditions, even the very extreme. Microorganisms that live in Yellowstone National Park's Grand Prismatic Spring are exposed to temperatures that can reach 70°C [158°F], and some bacteria can survive high doses of radiation or exposure to the vacuum of space. The following table is a collection of the extreme conditions that can support organisms on Earth.

	MINIMUM	MAXIMUM
Temperature	–20 °C	121 °C
Pressure	7 millibars	1.1 million millibars
pH	–0.06	11
Ionized Radiation	—	6,000 grays
Salinity	2 %	30 %

VERY RESISTANT ORGANISMS

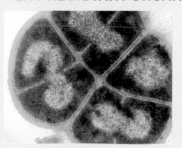

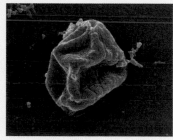

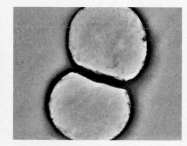

DEINOCOCCUS RADIODURANS
This bacterium has a way of repairing its DNA, making it both one of the most radiation-resistant organisms and mostly impervious to acids, cold, dehydration, and vacuums.

PYROLOBUS FUMARII
This species of bacteria lives near vents in ocean floors and can breed at temperatures up to 113°C [235°F]. The single-celled microbe *Geogemma barossii* is an even hardier species, surviving at temperatures up to 121°C [250°F].

PICROPHILUS
This genus of microorganisms includes two species of acidophiles, organisms that develop in acidic environments, capable of surviving a pH of −0.06—the lowest at which life can exist [in theory].

1

2

3

THE SEARCH FOR A NEW EARTH

Since the discovery of the first exoplanet, scientists has searched far and wide for a world that might have life.

Proxima b

1 This planet, also known as Proxima Centauri b, orbits the closest star to the sun, a red dwarf called Proxima Centauri found a little more than four light-years from Earth. Discovered in 2016, it has a mass slightly higher than that of our planet and its planetary equilibrium temperature is some −38 °C (−36.4°F).

Gliese 667Cc

2 This exoplanet orbits the star Gliese 667C, a red dwarf found 23.6 light-years from our solar system. It is a super-Earth (3.8 times Earth's mass), and probably has a radius much larger than our planet but smaller than the outer bodies of the solar system. Its equilibrium temperature is around 4°C (39.2°F).

TRAPPIST-1f

3 At some 39 light-years from Earth, the ultracold red dwarf star TRAPPIST-1 has seven rocky companion planets, three of which are situated in its habitable zone. TRAPPIST-1f is somewhat larger than Earth, but it has a significantly smaller mass (0.68 of Earth's mass) and a very cold equilibrium temperature, which measures slightly less than −50°C (−58°F).

Kepler-62f

4 Of the five rocky planets that make up the Kepler-62 system, this one orbits farthest from its star and is some 1,200 light-years away from us. Unless it has an atmosphere dense enough to retain heat from its sun, it is estimated to have a temperature of around −65°C (−85°F). If this is true, its climate is more similar to Mars's than to Earth's.

Kepler-186f

5 Found some 558 light-years away from Earth, this planet's radius is only 17 percent larger than Earth's. We don't precisely know its mass or composition, but like Kepler-62f it is likely quite a cold world (−85°C/−121°F) unless it has a very dense atmosphere.

Ross 128 b

6 This planet is found in the habitable zone of a red dwarf very close to the solar system (just 11 light-years away). With a mass around 1.4 times Earth's and a surface temperature that could be between −60°C (−76°F) and 21°C (69.8°F), it's a serious candidate in the search for life. It was discovered by the observatory in La Silla, Chile, in 2017.

THE FIRST EXOPLANET ORBITING A STAR

Exoplanet 51 Pegasi b was discovered in 1995. This planet, which is much larger than Earth, orbits 51 Pegasi, a small star located 50 light-years away.

HD 114762 b was discovered in 1989. It was the first exoplanet ever found, but it wasn't confirmed as one until three years later. The search for exoplanets heated up on October 6, 1995, when a gas giant was found outside the solar system orbiting

51 Pegasi (a sun-like star in the Pegasus constellation, about 50 light-years from Earth). It was named 51 Pegasi b, and scientists estimated that it had about half the mass of Jupiter.

In Numbers	DIAMETER	MASS	AVERAGE DISTANCE TO ITS STAR
 **Earth**	 12,742 km (7,918 mi)	5.972×10^{24} kg	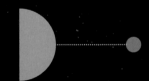 150 million km (93.2 million mi)
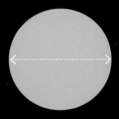 **51 Pegasi b**	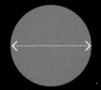 ≥ 140,000 km (86,992 mi)	 8.9215×10^{26} kg (mínimum)	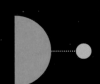 7.9 million km (4.9 million mi)

SIMILAR TO JUPITER

Exoplanet 51 Pegasi b's mass is at least half that of Jupiter, but it is probably the same size as, or larger than, the giant planet in the solar system.

A Small Star in the Pegasus Constellation

Located in the Pegasus constellation is 51 Pegasi, the star that 51 Pegasi b orbits. Its characteristics seem like they would be perfect for creating a habitable zone similar to the sun's.

ORBITAL PERIOD	AVERAGE TEMPERATURE	ORBITAL ECCENTRICITY	ORBITAL SPEED
			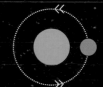
365.25 Earth days	14°C (57°F)	0.17	29.8 km/s (18.52 mi/s)
			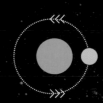
4.23 Earth days	1200°C (2192°F)	0	136 km/s (84.5 mi/s)

HD 189733B,
A BLUE GIANT

With a mass slightly greater than Jupiter's and orbiting close to its star, the gas giant HD 189733b is one of the largest known exoplanets.

One of the most studied giant planets, HD 189733b was first observed in 2005 when it passed in front of its star. It is not a normal gas giant: it has a mass 13 percent larger than Jupiter's and is located only 4.5 million kilometers (2.8 million mi) away from its star, but it travels at 152 kilometers (94.4 mi) per second, giving it an orbit of only 2.2 days. The extreme closeness of HD 189733b and other Jupiter-type planets to their stars (Earth, by comparison, is 150 million km/93.2 million mi from the sun) has led scientists to reexamine our theory of planetary formation. Before HD 189733b, the scientific consensus was that gas giants formed far from their stars, where low temperatures caused the compression of large amounts of gas around large, rocky nuclei.

Near or Far
There are only two theories that could explain the existence of HD 189733b and other gas giants orbiting close to their stars: they either formed very close to their star, contrary to what was originally believed, or they formed far away and have moved toward their star over time. The second theory is the one more generally accepted.

WATER EVERYWHERE

In 2007 the Spitzer Space Telescope detected water vapor in the atmosphere of HD 189733b, verified by Hubble a year later. To find it, astronomers analyzed changes in infrared light from its star as the planet slipped by in an infrared eclipse, filtering starlight through its outer atmosphere. While looking through an infrared filter, they noticed that for each wavelength the planet absorbed a different amount of light. This phenomenon could only be explained by the presence of one molecule: water.

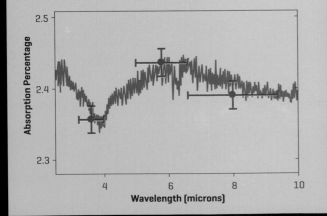

INTENSE BLUE
By measuring light as it passed behind its parent star, scientists made HD 189733b the first exoplanet to be observed in living color. Scientists deduced that it has tiny silicate particles in its atmosphere, blown around by howling winds. Its cobalt blue comes from glass raining down on the planet, with hazy clouds that scatter blue light.

Mapping the Atmosphere

Thanks to its size, and the fact that it's only 63.4 light-years from us, astronomers were able to study HD 189733b's atmosphere in greater detail than other bodies outside the solar system. They determined the atmosphere's temperature profile by observing it over a period of 33 hours, then created the first map of an exoplanet. It illustrates temperature variations: the lighter colors correspond to hotter zones.

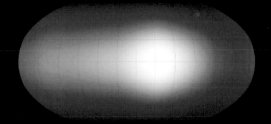

Two-Faced Temperature

HD 189733b is tidally locked with its parent star, which means the exoplanet only ever shows its star one face. The side that basks in constant starlight reaches surface temperatures of between 700°C and 940°C (1,292–1,724°F), while the upper layers of its atmosphere can reach several thousand degrees. Temperature differences between the two sides means its atmosphere has winds of up to 8,700 kilometers (5,400+ mi) per hour.

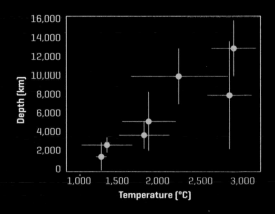

THE MILKY

IN

WAY

THE COSMOS

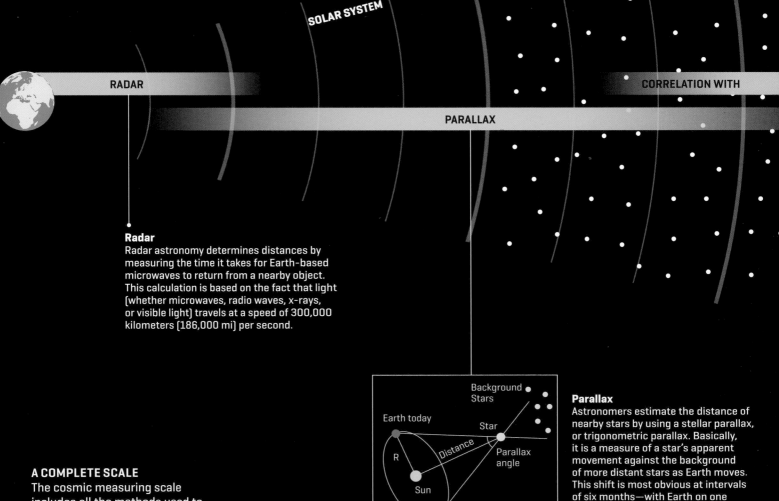

SOLAR SYSTEM

RADAR

CORRELATION WITH

PARALLAX

Radar
Radar astronomy determines distances by measuring the time it takes for Earth-based microwaves to return from a nearby object. This calculation is based on the fact that light (whether microwaves, radio waves, x-rays, or visible light) travels at a speed of 300,000 kilometers (186,000 mi) per second.

A COMPLETE SCALE
The cosmic measuring scale includes all the methods used to estimate distances between stars. The illustration represents some of the most important.

Background Stars

Earth today

Star

Distance

R

Parallax angle

Sun

Earth 6 months afterward

Parallax
Astronomers estimate the distance of nearby stars by using a stellar parallax, or trigonometric parallax. Basically, it is a measure of a star's apparent movement against the background of more distant stars as Earth moves. This shift is most obvious at intervals of six months—with Earth on one side of its orbit, and then six months later on the other—giving a baseline measurement that the star can be triangulated against.

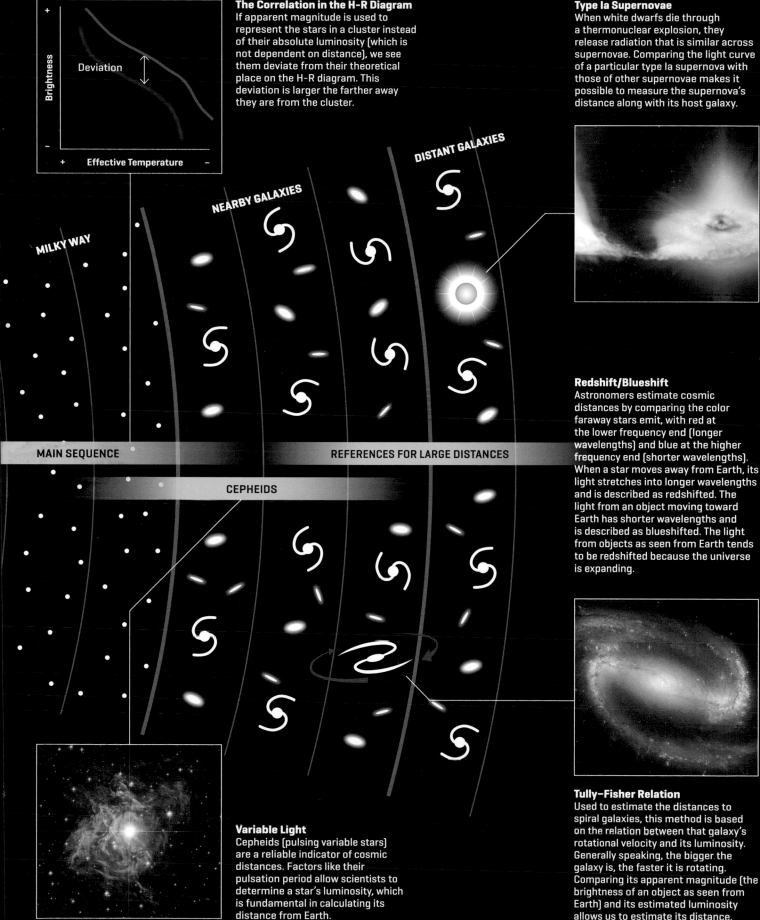

The Correlation in the H-R Diagram
If apparent magnitude is used to represent the stars in a cluster instead of their absolute luminosity (which is not dependent on distance), we see them deviate from their theoretical place on the H-R diagram. This deviation is larger the farther away they are from the cluster.

Brightness

Deviation

Effective Temperature

Type Ia Supernovae
When white dwarfs die through a thermonuclear explosion, they release radiation that is similar across supernovae. Comparing the light curve of a particular type Ia supernova with those of other supernovae makes it possible to measure the supernova's distance along with its host galaxy.

DISTANT GALAXIES

NEARBY GALAXIES

MILKY WAY

MAIN SEQUENCE

REFERENCES FOR LARGE DISTANCES

CEPHEIDS

Redshift/Blueshift
Astronomers estimate cosmic distances by comparing the color faraway stars emit, with red at the lower frequency end (longer wavelengths) and blue at the higher frequency end (shorter wavelengths). When a star moves away from Earth, its light stretches into longer wavelengths and is described as redshifted. The light from an object moving toward Earth has shorter wavelengths and is described as blueshifted. The light from objects as seen from Earth tends to be redshifted because the universe is expanding.

Variable Light
Cepheids (pulsing variable stars) are a reliable indicator of cosmic distances. Factors like their pulsation period allow scientists to determine a star's luminosity, which is fundamental in calculating its distance from Earth.

Tully–Fisher Relation
Used to estimate the distances to spiral galaxies, this method is based on the relation between that galaxy's rotational velocity and its luminosity. Generally speaking, the bigger the galaxy is, the faster it is rotating. Comparing its apparent magnitude (the brightness of an object as seen from Earth) and its estimated luminosity allows us to estimate its distance.

THE MILKY WAY IN THE KNOWN UNIVERSE

We used to think the Milky Way was at the center of the universe, but the more we learn the farther it moves away from it. This map of the observable universe shows that the Milky Way is only one in a web of galaxies.

Until the beginning of the 20th century, astronomers weren't sure if there were any galaxies besides the Milky Way. Today, the number of visible galaxies is estimated at over one trillion. Just like with ordinary matter, their distribution shows that the universe is uniform when using a scale of hundreds of millions of light-years. But when measured using smaller scales, galaxies are grouped in clusters and superclusters, all located amid a web of low-density gas filaments.

Groups, Clusters, and Superclusters

The enormous gravity of the galaxies means they tend to group together in structures ranging from only a few tens of galaxies, like the Local Group that includes the Milky Way, to clusters of hundreds. Regardless of how many clusters they have, the structures are all spherical and measure some 10 million light-years in diameter. All the galaxies orbit a massive center. Intergalactic space is not empty; it is occupied by hot gases that come from galaxies. These gases can escape one cluster's gravitational field and be attracted to another. Clusters can also orbit other, very massive ones known as superclusters; these groups are connected by filament structures that create a three-dimensional web. Our Local Group is part of the Virgo supercluster, or local supercluster, whose interior has a large center of mass: the Virgo cluster.

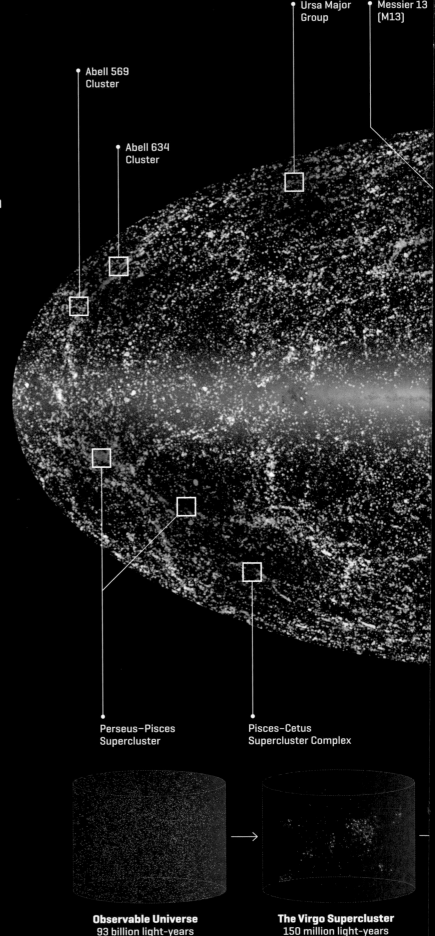

Abell 569
Cluster

Abell 634
Cluster

Ursa Major
Group

Messier 13
[M13]

Perseus–Pisces
Supercluster

Pisces–Cetus
Supercluster Complex

Observable Universe
93 billion light-years

The Virgo Supercluster
150 million light-years

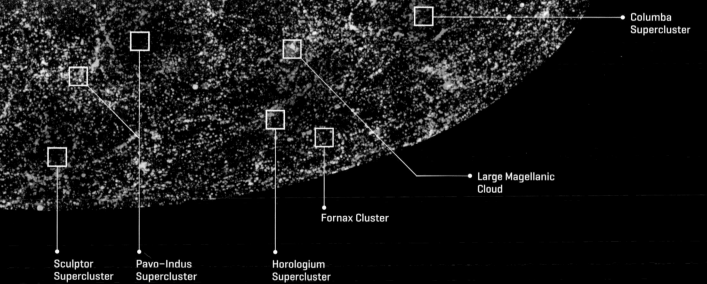

Corona Borealis
Cluster

Boötes
Supercluster

Coma Cluster

Virgo
Supercluster

Shapley
Supercluster

Ophiuchus
Cluster

Leo Cluster

Centaurus
Cluster

Hydra
Cluster

Center of the Milky Way

Norma and
the Great
Attractor

Columba
Supercluster

Large Magellanic
Cloud

Fornax Cluster

Sculptor
Supercluster

Pavo–Indus
Supercluster

Horologium
Supercluster

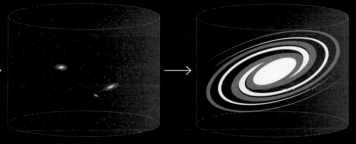

The Local Group
10 million light-years

Milky Way
100,000 light-years

THE SKY FROM THE EARTH
This projection of the celestial sphere as
seen from Earth, which uses the plane of
the Milky Way as a reference, shows the
major structures of the local universe.
The redder tones indicate large distances
between galaxies, while bluer ones are the
closest. The green ones are in between.

THE VIRGO SUPERCLUSTER

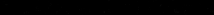

This supercluster includes several groups of galaxies, including the Milky Way. This graph shows where they can be found with respect to the galactic plane and the distance that separates them from our galaxy.

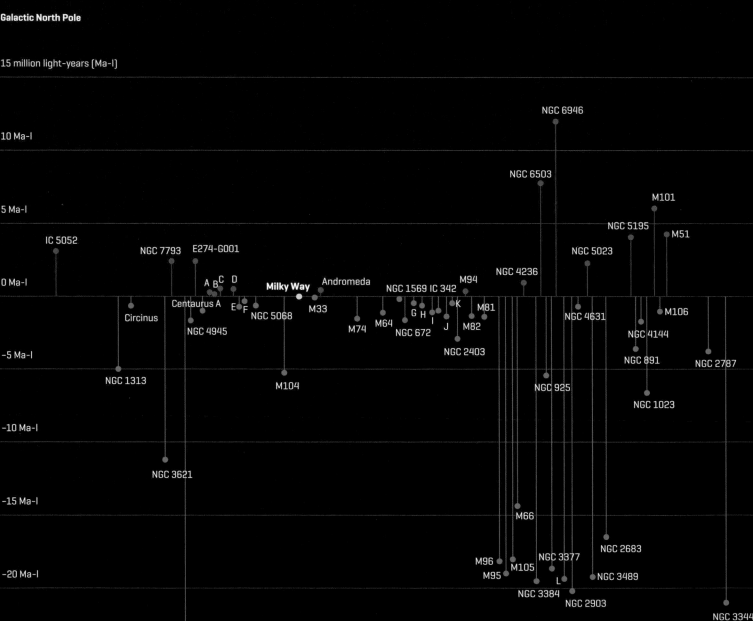

Galactic North Pole

15 million light-years [Ma-l]

10 Ma-l

5 Ma-l

0 Ma-l

−5 Ma-l

−10 Ma-l

−15 Ma-l

−20 Ma-l

IC 5052
NGC 7793
E274-G001
Centaurus A
Circinus
A B C D
E F
NGC 5068
Milky Way
M33
Andromeda
NGC 1569 IC 342
M94
NGC 4236
NGC 5023
NGC 5195
M101
M51
NGC 6503
NGC 6946
NGC 4945
NGC 1313
M104
M74
M64
NGC 672
G H
I
J
K
M82
M81
NGC 2403
NGC 925
NGC 4631
NGC 4144
NGC 891
M106
NGC 2787
NGC 1023
NGC 3621
M66
M96
M95
M105
NGC 3384
NGC 3377
L
NGC 2903
NGC 3489
NGC 2683
NGC 3344

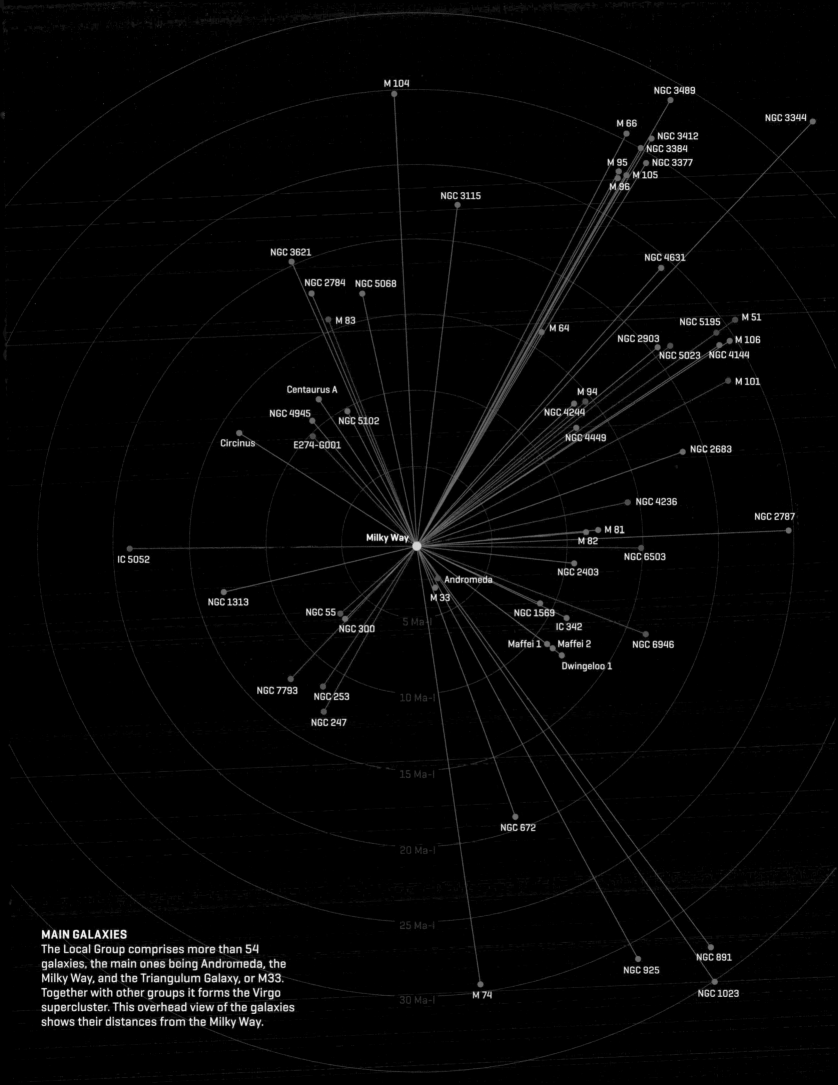

MAIN GALAXIES
The Local Group comprises more than 54
galaxies, the main ones being Andromeda, the
Milky Way, and the Triangulum Galaxy, or M33.
Together with other groups it forms the Virgo
supercluster. This overhead view of the galaxies
shows their distances from the Milky Way.

M 104
NGC 3489
NGC 3344
M 66
NGC 3412
NGC 3384
M 95
NGC 3377
NGC 3115
M 105
M 96
NGC 3621
NGC 4631
NGC 2784   NGC 5068
M 83
NGC 5195   M 51
NGC 2903   M 106
M 64
NGC 5023   NGC 4144
Centaurus A
M 101
NGC 4945
M 94
NGC 5102
NGC 4244
Circinus
E274-G001
NGC 4449
NGC 2683
NGC 4236
NGC 2787
M 81
IC 5052
M 82
NGC 6503
NGC 2403
NGC 1313
Andromeda
M 33
5 Ma-l
NGC 55
NGC 1569
NGC 300
IC 342
Maffei 1   Maffei 2   NGC 6946
Dwingeloo 1
NGC 7793
10 Ma-l
NGC 253
NGC 247
15 Ma-l
NGC 672
20 Ma-l
Milky Way
25 Ma-l
NGC 891
NGC 925
NGC 1023
M 74
30 Ma-l

other galaxies, from dwarf ellipses to irregulars, orbit these main galaxies.

Our Galactic Neighbors

The Canis Major dwarf galaxy is considered the closest galaxy to the Milky Way, although some experts say it's the remains of a galaxy absorbed by ours, making the Sagittarius dwarf elliptical galaxy some 70,000 light-years away the closest. The most prominent neighboring galaxies are the very bright Magellanic clouds, which are visible from the southern hemisphere. The intense gravitational pull the Milky Way exerts on its immediate neighbors allows it to siphon material from them.

5 | Millions of light-years from the center

4

3

LANIAKEA

In 2014 a group of astronomers defined the Virgo supercluster as part of a colossal supercluster called the Laniakea (Hawaiian for "immeasurable heaven"). This newfound supercluster may be part of a still larger structure that has not yet been defined. This illustration shows galaxies grouped together to form filaments, weaving the observable universe into a large cosmic web.

Milky Way

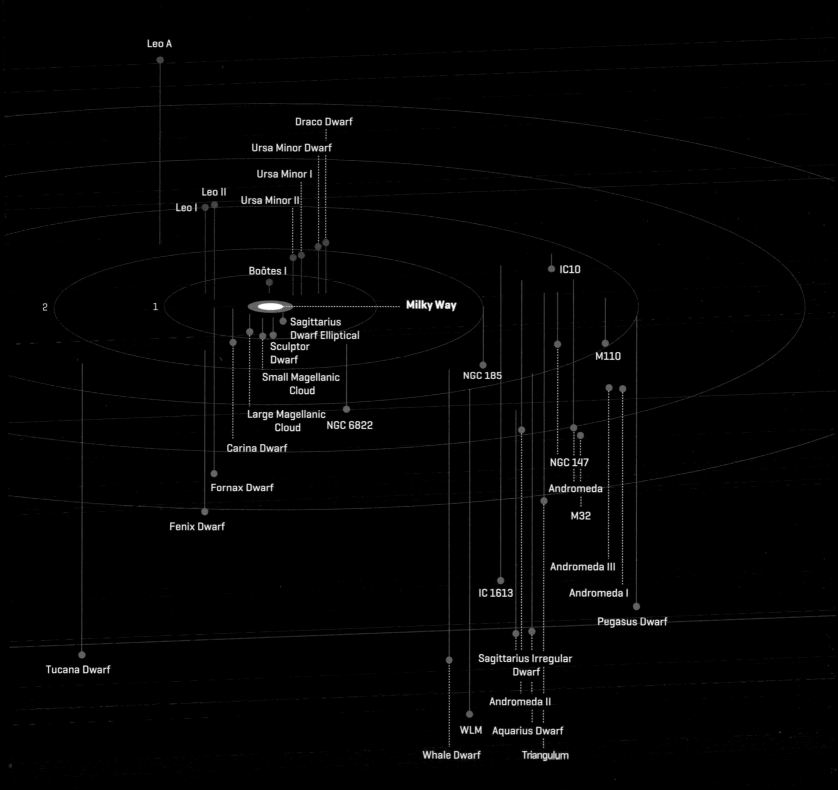

Leo A

Draco Dwarf

Ursa Minor Dwarf

Ursa Minor I

Leo II

Ursa Minor II

Leo I

IC10

Boötes I

Milky Way

2

1

Sagittarius
Dwarf Elliptical

Sculptor
Dwarf

M110

Small Magellanic
Cloud

NGC 185

Large Magellanic
Cloud   NGC 6822

Carina Dwarf

NGC 147

Fornax Dwarf

Andromeda

M32

Fenix Dwarf

Andromeda III

IC 1613   Andromeda I

Pegasus Dwarf

Tucana Dwarf

Sagittarius Irregular
Dwarf

Andromeda II

WLM   Aquarius Dwarf

Whale Dwarf   Triangulum

MAP OF THE LOCAL GROUP
This graphic illustrates the positions of
the Local Group galaxies in relation to the
Milky Way. The red peaks show the location
of structures found above the Milky Way's
plane, while blue dips show those found
below it.

THE DISCOVERY OF NEIGHBORING GALAXIES

Until the 1920s, we thought that nebulae—those faint objects riddled with filaments —were part of the Milky Way, but studying Cepheid stars in Andromeda and other nearby galaxies revealed their extragalactic characteristics.

After the Milky Way, the first recorded galaxy studied was Andromeda, observed in 961, although it's certain that prehistoric people in the Southern Hemisphere had already seen the Magellanic clouds. The next galaxies weren't observed until the 17th century, when telescopes reached a powerful enough magnitude. One of the first astronomical catalogs from that time, which included 110 astronomical objects (30 of which were galaxies), was compiled by French astronomer and comet hunter Charles Messier.

Viewing the Galaxies

During the 19th century, huge advances in telescopes allowed the discovery of thousands of galaxies, although it took another century and the ability to precisely measure wavelengths to figure out that these structures were outside the Milky Way. One of the scientists who contributed the most to the study of galaxies was American astronomer Edwin Hubble, who established the distances that separated the galaxies thanks to his study of Cepheid stars.

VISIBLE LIGHT
Different frequencies in the electromagnetic spectrum reveal different information about galaxies— in this case, Andromeda. Observed by the visible light portion of the spectrum, the brightness of stars appears dimmed due to clouds of stellar dust that cover the arms.

INFRARED
Young, hot stars heat up the clouds of dust that surround them. The dust shines red in this infrared image, revealing the true structure of the Andromeda galaxy.

MEASURING DISTANCE WITH CHANGING STARS

Cepheids are stars whose temperature and diameter shift regularly, changing their luminosity, while their pulsation period remains constant. If the pulsation period of a Cepheid is known, its absolute luminosity and distance can be calculated.

The Size of Galaxies

As the largest galaxy in the Local Group, Andromeda is the farthest object we can see with the naked eye. But its size is tiny compared to M87 and especially IC 1101, the largest known galaxy. These are found in the center of galactic clusters, surrounded by smaller galaxies that they steadily eat away at whenever they get close.

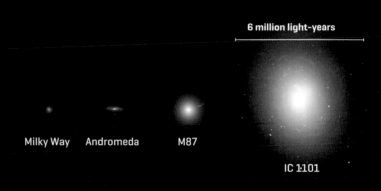

6 million light-years

Milky Way   Andromeda   M87

IC 1101

INFRARED AND X-RAY
This image combines far infrared with x-rays. While infrared shows the reservoirs of gas (in red) where new stars form, x-rays show older stars (in blue) concentrated in the galactic bulge and halo.

ULTRAVIOLET
This image of Andromeda shows off its young, hot stars along with dense stellar clusters.

THE MAGELLANIC CLOUDS

The Magellanic clouds are two dwarf galaxies near the Milky Way. Their closeness causes a gravitational interaction, distorting both the clouds and our own galaxy's disc.

Under favorable conditions, the Magellanic clouds can be observed with the naked eye near the celestial south pole. The larger of the two is known as the Large Magellanic Cloud and the smaller as the Small Magellanic Cloud. For the past few decades, astronomers have wondered whether they both orbit the Milky Way, although there are signs that the smaller cloud actually circles the larger one. Their high radial speed (or their velocity in the direction of an observer) shows that they are coming closer to our galaxy, which is deforming them

LARGE MAGELLANIC CLOUD
Found about 160,000 light-years from Earth, it is the third closest galaxy to the Milky Way and contains some 30 billion stars.

SN1987A
Seen in 1987 near the Tarantula Nebula, these remains of a supernova explosion have been one of the brightest observable events of the last 400 years.

Tarantula Nebula
This nebula is an active star formation region in the Local Group. Its luminosity is such that if it lived in the same galactic neighborhood as the closest nebulae in the Milky Way, it would cast shadows on Earth's surface.

NGC 1783
One of the brightest globular clusters in the Large Magellanic Cloud, its most significant characteristic is its youth: it is only 1.5 billion years old.

NGC 2014 and NGC 2020
Their shape comes from strong winds originating from recently formed, very hot stars. NGC 2014, a reddish color, is made almost entirely of hydrogen. NGC 2020's blue tint is due to oxygen

quite noticeably. Scientists are uncertain about their mass and speculate that they may have large halos of dark matter.

Clouds of Gas

The Magellanic clouds, especially the Large Magellanic Cloud, are active stellar factories, as evidenced by the large number of clusters and nebulae within them. Their significant gravitational attraction is relatively recent; without it, they would long ago have given the majority of their interstellar gas to the Milky Way.

MAGELLANIC STREAM

A stream of gas has been detected between the Magellanic clouds and the Milky Way, most of which was stolen from the Small Magellanic Cloud only two billion years ago. In this image, which combines radio waves and visible light, the gas current is shown in pink.

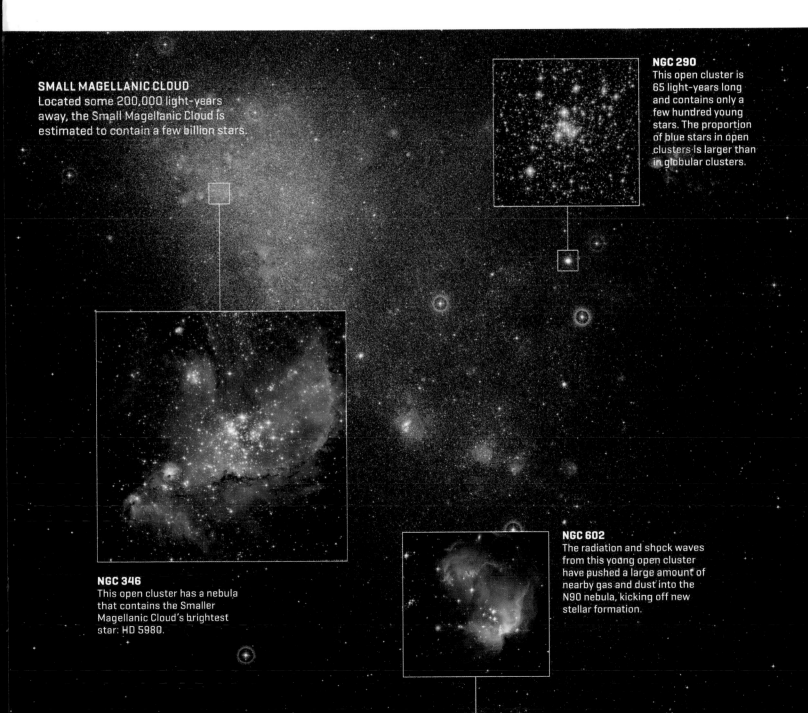

SMALL MAGELLANIC CLOUD
Located some 200,000 light-years away, the Small Magellanic Cloud is estimated to contain a few billion stars.

NGC 290
This open cluster is 65 light-years long and contains only a few hundred young stars. The proportion of blue stars in open clusters is larger than in globular clusters.

NGC 346
This open cluster has a nebula that contains the Smaller Magellanic Cloud's brightest star: HD 5980.

NGC 602
The radiation and shock waves from this young open cluster have pushed a large amount of nearby gas and dust into the N90 nebula, kicking off new stellar formation.

OTHER NEARBY GALAXIES

Andromeda Galaxy (M31)

1 This spiral galaxy, located 2.5 million light-years from Earth, is one of the largest and brightest in the Local Group.

NGC 3109

2 This irregular dwarf galaxy, some have theorized, may actually be a small spiral galaxy. Found some 4.2 million light-years away, it is outside the Local Group and seems to be moving away at a faster speed than originally thought.

IC10

3 Located in the Local Group approximately 1.8 million light-years from Earth, this dwarf galaxy is similar to the Small Magellanic Cloud in its characteristics and dimensions, except its star formation activity is even greater.

Barnard's Galaxy

4 This irregular galaxy is some 1.6 million light-years away and part of the Local Group. Similar in structure and composition to the Small Magellanic Cloud, it has been difficult to study because of how close it lies to the Milky Way's plane.

M32

5 A dwarf elliptical galaxy found some 2.65 million light-years away, this satellite of Andromeda experiences enormous gravitational forces from its giant neighbor.

Wolf–Lundmark–Melotte Galaxy

6 This irregular galaxy, found some three million light-years away, is very isolated from the others within the Local Group. It has an elongated shape and is larger than the group's other dwarf galaxies.

Triangulum Galaxy (M33)

7 This is the third largest galaxy in the Local Group, after Andromeda and the Milky Way. Like them it has a spiral structure, although fewer stars, and it is believed to be linked gravitationally to Andromeda.

THE COLLISION WITH ANDROMEDA

Thanks to observing many galaxies as they evolve, we know that the Milky Way is a complex and changing galaxy and that it will eventually fuse with the largest of its neighbors, Andromeda.

Nearby galaxies interact with each other, causing them to distort as they exchange gas and dust. If they are near enough, they can cross paths in what is known as a galactic collision. The stars that make up the colliding galaxies usually do not collide; instead, their interstellar dust interacts to create flares and diverse structures, such as bars and rings. When galaxies pass by but don't go through each other, they fuse, combining into a larger structure.

Joining Andromeda

Since birth, the Milky Way has been increasing in size by fusing with other galaxies, although this has not happened in the last billion years. It is currently increasing its mass by stealing gas from the Magellanic clouds. The Milky Way will eventually run parallel with its giant neighbor, the Andromeda galaxy. Both are moving closer together at 300 kilometers (186 mi) per second and are expected to collide, forming a giant elliptical galaxy, some seven billion years from now.

The Fusion of Galaxies

Larger galaxies are formed when smaller ones fuse. Spirals are the result of smaller galaxies fusing, while ellipticals are created by larger galaxies colliding.

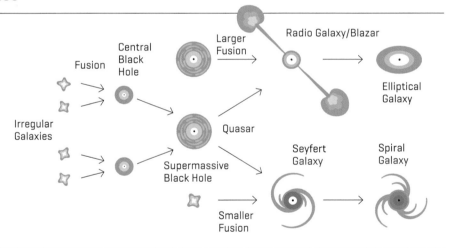

Irregular Galaxies

Fusion

Central Black Hole

Larger Fusion

Radio Galaxy/Blazar

Elliptical Galaxy

Quasar

Supermassive Black Hole

Seyfert Galaxy

Spiral Galaxy

Smaller Fusion

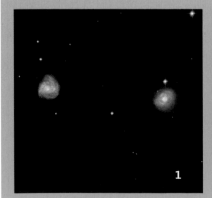

1

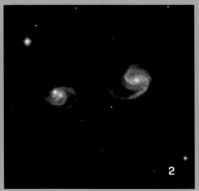

2

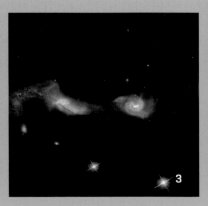

3

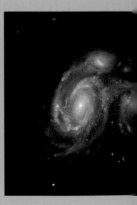

A NEW GALACTIC ORDER

This illustration imagines the first interaction between the Milky Way and Andromeda. From a hypothetical planet, the sky would look as if it were on fire because of all the emission nebulae, which would also mean a high rate of stellar formation.

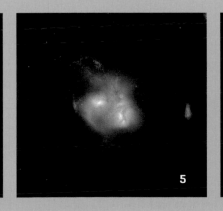

5

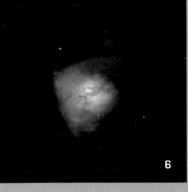

6

BIRTH OF AN ELLIPTICAL GALAXY

According to the most widely accepted theories, large elliptical galaxies come from the fusion of spiral galaxies. One of the first signs of fusion is the exchange of materials [1], which creates the so-called tidal tails [2] that sometimes remain after the interaction [3]. The proximity of the cores causes enormous distortions [4] and captures streams of material, resulting in a burst of stellar formation [5]. Some galaxies have bright stellar clusters made by such dramatic explosions [6].

THE FUTURE OF THE MILKY WAY
Within some seven billion years, the Milky Way and Andromeda will fuse to form a colossal elliptical galaxy. With almost a billion stars, the resulting galaxy will dominate this corner of the universe. This image illustrates how the night sky could look from a hypothetical galaxy within this new galactic landscape.

GLOSSARY

GLOSSARY

A

ACCRETION The accumulation of material, normally gases, that happens around stars, planets, and other celestial objects because of gravity.

ALBEDO Quantity of radiation reflected by an object based on how much it received.

ANGULAR MOMENTUM Physical magnitude that indicates the movement of a rotating object. Angular momentum can be divided into its components: the orbital angular momentum, or its movement around an external axis, and its rotation around its own axis.

ANTIMATTER Dark particles that have equivalents with ordinary matter. Although both have the same mass, they have opposite electrical charges. The positron is the opposite of an electron (with a negative charge), and the antiproton is the opposite of a proton. When ordinary matter and antimatter come into contact they annihilate each other, making energy.

APHELION Farthest point in the orbit of a celestial body around the sun.

ASTEROIDS Small objects in the solar system, some of which are considered minor planets. Asteroids are believed to be the remains of ancient planetesimals.

ASTEROSEISMOLOGY Branch of astronomy that studies the internal structure of stars from their oscillations. This same branch, when applied to the sun, is known as helioseismology.

ASTROMETRY Branch of astronomy dedicated to precise measurements of positions and movements of stars and other celestial bodies.

ATOM Smallest component of ordinary matter that has properties of a chemical element.

ATOMS, HEAVY Atom that contains a larger number of protons and neutrons than iron, whose atomic number is 26.

AURORA Natural light visible in polar regions of a planet or moon. The product of the interaction of solar wind with the magnetosphere, which causes charged particles to fall into the upper layers of the atmosphere.

AXIS OF ROTATION Imaginary line about which a celestial object rotates.

B

BELT, ASTEROID Disc of material that surrounds the sun found between the orbits of Mars and Jupiter. It has a multitude of asteroids, and its mass is equivalent to 4 percent of the mass of the moon.

BELT, KUIPER Disc of material surrounding the sun and found outside the orbit of Neptune. Although it looks like the asteroid belt, it is 20 times wider and has a mass 200 times larger.

BIG BANG Name given to the theory of the origin of the universe, which was created some 13.8 billion years ago by the rapid expansion from a point of infinite density.

BLACK HOLE A region of space produced by a dense body, with an intense gravitational pull from which nothing can escape.

BLACK HOLE, STELLAR Black hole created by the collapse of a very massive star at the end of its life.

BLACK HOLE, SUPERMASSIVE The largest black holes in the universe. Although the formation process is still a mystery, they could evolve from a stellar black hole that accumulates material.

BLAZAR Compact quasar associated with a supermassive black hole found in the center of a giant elliptical galaxy. Blazars can be seen when a stream of material emitted by the quasar heads toward Earth.

BRIGHTNESS Amount of luminosity emitted by celestial objects according to the point of view of an observer on Earth.

C

CENTER OF MASS Also called the barycenter, it indicates the place where two or more objects orbit each other and is the point about which the bodies orbit. When the difference between the masses of the bodies is very large, the barycenter is found almost in the center of the more massive one. See also Star, binary.

CEPHEID Type of variable star that regularly changes its diameter and temperature, which causes changes in its brightness over a stable period based on its intrinsic luminosity.

CHANDRASEKHAR LIMIT Maximum mass that a stable white dwarf can have. The accepted limit is currently 1.4 solar masses. If a white dwarf is above this limit, it can become a neutron star or a black hole.

CHROMOSPHERE The second layer in the sun's atmosphere just above the photosphere and below the transition region.

CLOUD, MOLECULAR Interstellar cloud with enough density and size to allow for molecular hydrogen (H_2) to form. Its interior has denser regions that, if gravity is intense enough, trigger the star formation process.

CLOUD, OORT Cloud that surrounds the solar system, but whose existence has not been proven definitively. Its outer border is the edge of the solar system.

CLUSTER, GALACTIC Structure formed by hundreds or thousands of gravitationally linked galaxies. They are the largest structures in the universe linked by gravity and can join to form superclusters.

CLUSTER, STELLAR Group of stars linked by gravity. There are two large groups: globular clusters, giant groups of very old stars; and open clusters, which have a few dozen young stars.

CNO (CARBON–NITROGEN–OXYGEN) CYCLE One of two fusion mechanisms, together with the proton-proton chain, that stars use to convert hydrogen into helium. This is the dominant reaction in massive stars.

COMET Small frozen object that, when close to the sun, starts releasing gases from its surface. This process creates the so-called tail of material.

COMET NUCLEUS Solid part of a comet made of frozen rock, dust, and gases that, when it draws near to the sun, evaporates and creates its characteristic head and tail.

CONVECTION One of the forms of heat transfer. In astronomy, convection is responsible for energy transmission in a star, moving hot plasma outward and cold plasma inward.

CORE, GALACTIC The center of a galaxy, normally made up of a large group of stars. If the galaxy is significantly large enough, it can have a supermassive black hole.

CORE, PLANETARY Innermost layers of a planet. The planetary core can be made of solid or liquid layers.

CORE, SOLAR Hottest region of our star and the solar system. It has a temperature of some 15.5 million degrees Celsius (27 million degrees Fahrenheit).

CORONA Region of plasma that surrounds the sun and other stars. It extends millions of kilometers/miles and is noted for its high temperatures of over 1 million degrees Celsius (1.8 million degrees Fahrenheit).

COSMIC DISTANCE SCALE A group of methods used in astronomy to determine the distances between celestial objects.

CRUST Solid outermost layer of a rocky planet or moon. It should not be confused with the mantle, which has a different chemical composition. In Earth's case, it is divided into the oceanic and continental crusts.

CRYOVOLCANO Type of volcano that ejects volatile elements (water, ammonia, methane, etc.) from a rocky base.

D

DIFFERENTIAL ROTATION Phenomenon that occurs when different parts of a rotating object move at different angular speeds, revealing that the object is not solid.

DISC, ACCRETION Structure formed by a diffuse material that orbits a massive object, usually a star. The friction causes the material in the disc to fall toward the object.

DISC, GALACTIC Plane in which the arms and bars of spiral galaxies are found. It has the highest concentration of gas and dust in the galaxy, along with a large group of young stars.

DISC, PROTOPLANETARY Disc made of a dense layer of gas and dust that orbits a recently formed star. With time, the protoplanetary disc will give rise to different bodies in a star system.

DWARF, BLACK Theoretical stellar object that is defined as the result of a white dwarf that, after billions of years, has released the majority of its energy into space and does not emit light.

DWARF, BROWN Substellar object between a giant planet and a star. It is also called a "fallen star" because it did not gather enough material during its formation to start the fusion of hydrogen.

DWARF, RED A relatively small and cold main sequence star. It is the most common type of star in the Milky Way, noted for having an extremely long life that can last billions of years.

DWARF, WHITE The remains of a star; the core of an old star similar to the sun in the final stage of its life that has ejected its external layers into space.

E

ECLIPSE Astronomical phenomenon that is caused when an object is obscured temporarily. It happens when an object passes through the shadow of another object, as in the case of a lunar eclipse, or when a second object passes between the observer and what he or she is observing, which happens during a solar eclipse.

ECLIPTIC The term refers to two concepts. The first is the circular path that describes the sun's movement in the celestial sphere over a year. The second is the orbital plane in which Earth moves around the sun. It is used as a reference to indicate the orbital incline of other objects in the solar system.

ELECTROMAGNETIC SPECTRUM See Spectrum, electromagnetic.

ELECTRON Subatomic particle with a negative electrical charge. It is one of the elemental particles and is fundamental in many interactions (gravitational, electromagnetic, and weak) in nature. It also participates in nuclear reactions, such as the fusion process in the interior of a star.

ELEMENTS, HEAVY Group of elements, excluding hydrogen and helium, that formed during the big bang. They include all elements created inside stars as well as through other mechanisms, such as supernovae and collisions of neutron stars.

EQUINOX Point during the year at which night and day have approximately the same length. This happens twice a year on Earth: March 20 or 21 and September 22 or 23. This marks the start of spring and fall, respectively, in the Northern Hemisphere.

EVENT HORIZON Region of spacetime in which events cannot affect an external observer. The most popular example is that of black holes: the event horizon is the region in which gravity is so intense that not even light can escape.

EXOPLANET Also called extrasolar planet, it is a planet located outside the solar system that orbits a star.

EXTREMOPHILES Organisms that develop in extreme conditions, either physical or geochemical, which are harmful to most forms of life on Earth.

F

FACULA Latin for "little torch." Refers to a point on the surface of a celestial body that is noted for its brightness. On the surface of the sun, faculae are small areas of convection that last for several minutes.

FROST LINE Indicates the distance from a young central star where the temperature is sufficiently low to allow for volatile elements to exist as grains of ice.

G

GALACTIC ARMS Structures of stars, gas, and dust that extend from the center of spiral galaxies. They are the main characteristics of these galaxies.

GALACTIC BULGE Giant group of stars in a relatively compact space. It also refers to the visible bulge in the center of many galaxies.

GALACTIC HALO Approximately spherical region that extends beyond the visible part of the galaxy and is made mostly of low-density interstellar gas, old stars, and dark matter.

GALACTIC SUPERCLUSTERS Groups of galactic clusters, among the largest structures in the cosmos.

GALAXY   A group of celestial objects, such as stars, planets, nebulae, and black holes, linked gravitationally. Galaxies are classified according to their size and structure.

GALAXY, ACTIVE   Galaxy that has an active galactic core. Its center, or at least part of it, has a luminosity much higher than normal in its electromagnetic spectrum. This luminosity is due to the presence of a supermassive black hole.

GALAXY, BARRED SPIRAL   Occasionally, spiral galaxies can have a barlike structure at their core, as is the case with the Milky Way. This structure acts like a stellar nursery, attracting interstellar gas to the center of the galaxy.

GALAXY, RADIO   Exceptionally luminous galaxy in the radio spectrum.

GALAXY, SEYFERT   Galaxy with an active core, very similar to a quasar, but less luminous and with an observable galactic structure.

GALAXY, SPIRAL   Named for its appearance, a galaxy in which the stars and gas clouds are concentrated in one or more tightly wound spiral arms that spread outward from its center. Most spiral galaxies consist of a flat, rotating disc and a central concentration of stars, the bulge.

GAS   One of the four states of matter, found between the liquid and plasma states. It is different from liquids and solids due to the large separation between its component particles.

GIANT, RED   Giant low-mass star that is in the last stages of stellar evolution.

GRAVITATIONAL COLLAPSE   Mechanism by which a celestial object contracts on itself due to its own gravity.

GRAVITATIONAL COUPLING   Phenomenon that occurs from the interaction between two objects in orbit that causes the rotational period to be identical to the orbital period. Also called tidal locking.

GRAVITATIONAL FIELD   Model used to describe the influence of a massive object on the space around it, generating a force on other massive objects.

GRAVITATIONAL LENS   Distribution of material between a faraway source of light and a hypothetical observer that is capable of bending light from the source on its way to the observer.

H

HABITABILITY   Measure of the potential of a planet or moon to have hospitable environments for life, along with the capability to develop its own living organisms.

HEAD   Cloud of gas and dust that surrounds the core of a comet when it enters the inner solar system. Upon heating up, part of its material vaporizes, forming a thin atmosphere.

HELIOPAUSE   Region that marks the point at which the solar wind is slowed down by the interstellar medium.

HELIOSPHERE   Region of space in the form of a bubble that is dominated by the sun and extends beyond Pluto's orbit. It is fed by the solar wind, which maintains an active front against the pressure of the interstellar medium. It is made mostly of ions from the sun's atmosphere.

HI AND HII REGIONS   Clouds in the interstellar medium made of neutral atomic hydrogen (in the case of HI regions) or ionized atomic hydrogen (HII regions).

H-R DIAGRAM   Hertzsprung–Russell diagram, used to classify stars in relation to their luminosity and temperature, the latter of which also shows a star's color.

HYDROGEN, IONIZED   Protons in this state lack electrons. In the universe, they can be found in HII regions, clouds of partially ionized hydrogen, where star formation has recently taken place.

HYDROGEN, METALLIC   State of hydrogen in which it can act as an electrical conductor. At high temperatures and pressures, metallic hydrogen exists as a liquid rather than a solid.

HYDROSTATIC EQUILIBRIUM   In stars, equilibrium is maintained between the force of gravity, which attracts material toward its center, and nuclear reactions, which take place in the core and release large amounts of energy in the other direction. This equilibrium gives the star a spherical shape.

HYPERNOVA   Also known as a superluminous supernova, it is a type of stellar explosion that causes a luminosity 10 times greater than a standard supernova's.

I

INFRARED   Type of electromagnetic radiation with a wavelength greater than that of light, making it invisible to the human eye.

INTERSTELLAR DUST   Remains of ancient stars, made of microscopic particles that can be found in the space between star systems.

IONIZATION   Process in which a molecule dissociates into its ions or the conversion of an atom or molecule into an ion. An ion is an atom or group of atoms that has an electrical charge due to the gain or loss of one or more electrons.

IONOSPHERE   Upper part of Earth's atmosphere that is ionized by solar radiation.

J

JET   A stream of matter made of an ionized material (having an electrical charge) that accelerates to speeds near that of light.

K

KINETIC ENERGY   Energy that an object has due to its movement. It can be defined as work (force necessary to change the object's state) that is needed to accelerate an object of some mass from its resting state to a determined speed. The object conserves the energy obtained during its acceleration unless its speed changes.

L

LAGRANGE POINTS   Positions in the orbital configurations of large objects in which a third, smaller object can maintain its position relative to them by gravity alone.

LAVA STREAM   Layer of lava that a volcano emits when it erupts. The stream can descend the slope of a mountain or, if it is ejected through a fissure, create broad lava fields.

LIGHT CURVE   Graphic that shows the intensity of light from an object over a determined period of time.

LIGHT-YEAR   Distance traveled by light in a year, 9.46 trillion kilometers (5.88 trillion mi). It is one of the standard astronomical measurements.

LUMINOSITY   In astronomy, it is the amount of energy given off by a celestial object, such as a star or galaxy, in a certain interval of time.

M

MAGMA Mix of molten or semimolten rock and other elements found below the surface of Earth and other rocky planets and moons. It accumulates in magma chambers, which can form a volcano or solidify below the surface. When it emerges, it is known as lava.

MAGNETOSPHERE Region of space around a celestial object in which charged particles are dominated by the magnetic field of the object.

MAGNITUDE Measurement that indicates the brightness of an object at a specific wavelength. There are two types: apparent and absolute. Apparent magnitude is the brightness of an object as perceived from Earth. Absolute magnitude is the brightness of an object if it were observed at a standard distance of 10 parsecs (32.6 light-years) away from Earth. It is inversely proportional: the lower the magnitude, the more luminous the object.

MAIN SEQUENCE The main phase in the life of a star. The time a star remains in the main sequence is shorter the larger its mass.

MANTLE Internal layer between the crust and the core of a rocky planet or other rocky celestial body.

MASS Property of a physical object that indicates its resistance to acceleration when a force is applied. Also refers to its gravitational attractive force on other objects.

MATTER, DARK Type of matter, distinct from ordinary matter, that is thought to be responsible for 95 percent of the total mass of the universe. Although it cannot be observed directly, its existence allows us to explain many astronomical phenomena. It neither emits nor interacts with electromagnetic radiation.

MATTER, DEGENERATE Matter inside a star that forms at the end of its life.

MATTER, ORDINARY All the material that can be seen. Also known as baryonic matter.

METALLICITY Term used to describe when a star has elements heavier than helium. In astrophysics, all of these elements are known as metals.

METEORITE Portion of a meteoroid or an asteroid that hit the surface of some object in the solar system and was not destroyed.

MICROWAVES Form of electromagnetic radiation whose wavelength is between 1 meter (3.3 ft) and 1 millimeter (0.039 in). Radioastronomy allows us to study this type of radiation that comes from stars, planets, galaxies, and other celestial objects.

MOTION, PROGRADE In astronomy, motion of a celestial object in the same direction as other objects within its system. Also called direct motion.

MOTION, RETROGRADE In astronomy, motion of a celestial object in a direction opposite to that of other objects within its system, as observed from a particular vantage point, such as looking down on the object from the north.

N

NEBULA Latin for "cloud" or "fog." A giant cloud of dust and gases (hydrogen, helium, and other ionized gases) in space. Some nebulae are the product of dying stars; others, called star nurseries, are areas where new stars are beginning to form.

NEBULA, ABSORPTION see Nebula, dark

NEBULA, DARK Also known as absorption nebula, it has a high enough density to obscure the light from objects behind it and is cold enough at the same time to not emit visible light.

NEBULA, EMISSION Nebula formed of ionized gases that emits light of different colors depending on the ionization and composition of the gases.

NEBULA, PLANETARY Nebula made when a sunlike star reaches the end of its life and expels its external layers into the surrounding space. The star's core remnant lights up the nebula and causes it to look like an emission nebula.

NEBULA, REFLECTION Nebula that reflects light from nearby stars. The energy received is not sufficient to ionize gas, but it does cause the light to disperse and illuminate the particles that make up the nebula.

NEUTRINO Elemental particle that interacts only with gravity and the weak electromagnetic force. It is difficult to detect, as it goes through ordinary matter without any type of interaction.

NEUTRON Subatomic particle with no electrical charge. Its mass is very similar to that of a proton. Together, they form the nucleus of an atom.

NUCLEAR FUSION Process in which several atomic nuclei combine to form a new nucleus. The process can both release and absorb energy and is the production method for energy in stars.

O

OBLIQUITY Rotational axis incline of a plane with respect to the orbital plane of the system in which it lies. Also called axial tilt.

OLIGARCHY Term used in some planetary formation theories to define objects with a size that could be the same as the planets of the solar system.

ORBIT Curved trajectory that an object follows due to the gravitational influence of another.

ORBITAL ECCENTRICITY Parameter that indicates the deviation from the orbit of an object being a perfect circle. A value of 0 indicates a perfect circle, 0 to 1 indicates more elliptical orbits, and higher values equal to or greater than 1 indicate parabolic orbits.

ORBITAL INCLINE Measurement that indicates the deviation of a celestial object's orbit with respect to the plane of reference used in the star system.

ORBITAL RESONANCE Phenomenon that occurs when orbiting bodies exert regular, periodic gravitational influence on each other.

P

PARALLAX Apparent movement of an object in relation to more distant objects observed from two different viewpoints. In astronomy, it is used to measure the distance of faraway objects.

PERIHELION Closest point in the orbit of a celestial body around the sun.

PHOTON Elementary particle associated with all forms of electromagnetic radiation, from gamma rays to radio waves. Photons do not have a mass and travel through space at almost 300,000 kilometers (186,411 mi) per second.

PHOTOSPHERE Outermost layer of a star from which light is emitted.

PLANET, DWARF   Object with a planetary mass that is not a planet or a moon. It orbits a star and is spherical due to its own gravity. However, it does not have orbital dominance, sharing its orbit with other objects of a similar size (excluding its moons).

PLANET, GAS   Planet composed mostly of gases, such as hydrogen and helium, with a small rocky core. In the solar system, Jupiter, Saturn, Uranus, and Neptune are the gas planets.

PLANET, ROCKY   Planet composed mostly of rocks and metals. In the solar system, Mercury, Venus, Earth, and Mars are the rocky planets. Also known as a terrestrial planet.

PLANE, GALACTIC   Plane in which most of the mass of the galaxy can be found.

PLANE, INVARIABLE   Measure of all orbital planes and rotations of the objects that make up a planetary system.

PLANE, ORBITAL   Geometric plane in which the orbit of one object around another is found.

PLANETESIMAL   Small celestial object formed of dust, rock, and other material. In the planet formation theory, planetesimals collided, fused, and through gravity merged with other bodies to form the early solar system.

PLASMA   One of the four states of matter. In plasma, all atoms are ionized.

POLE, MAGNETIC   Place on the surface of an astronomical object in which the magnetic field lines are vertical. Generally, it is found on the axis of rotation of a celestial object.

POSITRON   Electron antiparticle (or antimatter). It has the same mass but a positive charge. When a positron and an electron collide, they destroy each other, emitting gamma-ray photons.

PROTON   Subatomic particle with a positive charge; it is a fundamental component of the nucleus of all atoms. The number of protons present in each atomic nucleus defines the properties of the element.

PROTON-PROTON CHAIN   One of two fusion mechanisms, together with the CNO cycle, that stars use to convert hydrogen into helium. This reaction is the main one in stars of the sun's size or smaller.

PROTOPLANET   Large planetary object that originated in a protoplanetary disc. Protoplanets form by the combination of planetesimals and are the precursors of planets.

PROTOSTAR   Young star that is still absorbing mass from its molecular cloud. It is in the first phase of stellar evolution, and the duration of this phase determines the size of the future star.

PULSAR   Neutron star that emits a stream of charged particles at very high speeds. The radiation emitted by this stream is visible when headed toward Earth.

Q

QUASAR   Type of active galaxy with an extreme luminosity, consisting of a supermassive black hole surrounded by an accretion disc.

R

RADIATION, ELECTROMAGNETIC   Radiation that is created by waves of the electromagnetic field: radio waves, microwaves, infrared, visible light, and x-rays, ultraviolet rays, and gamma rays.

RADIATION, IONIZING   Radiation that has enough energy to ionize matter, or separate electrons from molecules and atoms. x-rays, gamma rays, and ultraviolet rays are types of ionizing radiation.

RADIOACTIVITY   Physical phenomenon by which atomic nuclei of certain unstable chemical elements deteriorate into other, stable elements, emitting ionizing radiation.

RAYS, COSMIC   High-energy radiation from outside the solar system and the Milky Way. When they impact Earth's atmosphere, they cause a cascade of secondary particles.

RAYS, GAMMA   Range of the electromagnetic spectrum with the highest-energy radiation. They can be created as a product of radioactivity of an unstable atomic nucleus. Gamma rays are capable of ionizing other atoms, as well as causing mutations in living organisms.

RAYS, X-   Type of very energetic radiation, although less so than gamma rays. They are emitted by astronomical objects with temperatures above 1 million degrees Celsius (1.8 million degrees Fahrenheit).

REDSHIFT   Phenomenon that occurs when light emitted by a source moving away from the observer appears to stretch, causing a displacement toward red in the electromagnetic spectrum with respect to the original light emitted. The opposite, when light is shortened because the source moves toward the observer, is known as blueshift.

S

SHOCK WAVE   An area of very high pressure that moves faster than the speed of sound through a medium (e.g., air, water, or a solid substance). Throughout the universe, supersonic shock waves propel cosmic rays and supernova particles to velocities near the speed of light.

SINGULARITY   Point in space in which the laws of physics stop working. For example, the center of a black hole is a singularity, as it has infinite gravity.

SOLAR ATMOSPHERE   Outermost region of the sun composed of the photosphere, chromosphere, and corona. It can be seen with the naked eye from Earth during a solar eclipse.

SOLAR FLARE   A sudden, powerful burst of radiation from the release of magnetic energy that has built up in the solar atmosphere, usually above sunspots.

SOLSTICE   The time when the sun reaches its highest or lowest point in the sky at noon, resulting in the shortest (winter solstice) or longest (summer solstice) days in the year.

SPECTRUM, ELECTROMAGNETIC   Complete range of frequencies of electromagnetic radiation, extending from gamma waves, with the shortest wavelength (less than the size of an atom) to extremely low-frequency radio waves, with the longest wavelengths, and including visible light.

SPECTRUM, VISIBLE   Portion of the electromagnetic spectrum that is visible to the human eye. This corresponds to wavelengths between 390 and 700 nanometers. Also called visible light.

STAR, BINARY   A system of two stars orbiting a common center of mass (barycenter). To the naked eye, the two stars appear to be a single point of light.

STAR, GIANT Star with a radiance and luminosity much larger than that of the sun. A giant star has left its main sequence phase or is in the process of doing so, having consumed all of the hydrogen in its core.

STAR, NEUTRON Stellar remnant created when a star with a mass larger than eight to 25 times that of the sun reaches the final stage of its life. Due to its small diameter of tens of kilometers/miles, it is one of the densest objects in the universe.

STAR, SUPERGIANT Star with at least eight times the mass of the sun that has consumed all the hydrogen in its core. To do this, it rapidly begins fusing helium and other elements in its interior.

STAR, T TAURI Very young star that is in the last stages of its formation. Once finished, the star can start fusing hydrogen in its interior, starting the first phase of the main sequence.

STAR, VARIABLE A star that, seen from Earth, has a variable brightness. This change can be due to a change in the light emitted or something that partially blocks the light from our perspective.

STARS, POPULATION I, II, AND III System of star classification based on the number of elements present in their composition (heavier than hydrogen and helium) and their age. Population I stars are the youngest and have the highest number of heavy elements. Population II stars are the oldest and have lower amounts of heavier elements. Population III stars, which no longer exist, were created during the big bang and were made of only hydrogen and helium.

STELLAR WIND Stream of charged particles that come from the outer layers of a star. It is made up of electrons, protons, and other particles. In the sun's case, it is called solar wind.

STORM, GEOMAGNETIC Temporary disruption of Earth's magnetosphere caused by the solar wind or a magnetic field that interacts with our planet.

STORM, SOLAR Definition used as a reference for especially intense solar flares, as well as coronal mass ejections.

STRATOSPHERE Second thickest layer of Earth's atmosphere, above the troposphere and below the mesosphere. It contains about 20 percent of the planet's atmosphere.

SUB-EARTHS Planets that have much less mass than Earth or Venus. They are the hardest to detect in the search for exoplanets.

SUNSPOT Temporary phenomenon visible in the sun's photosphere. Easy to recognize because it appears darker than the surrounding regions. It follows a local increase in the sun's magnetic field that slows down convection (the circulation of its material).

SUPER-EARTH Planet with a mass greater than Earth's but less than that of Uranus or Neptune. Although a consensus has not been reached, generally the upper limit is thought to be 10 Earth masses.

SUPERGIANT, RED One of the largest types of stars (by volume) in the universe. One of the last stages of stars that are no longer in their main sequence.

SUPERNOVA Extremely energetic phenomenon that happens at the end of a massive star's life and causes its destruction, either fully or partially.

T

TACHOCLINE Region of transmission between the radiation and convection zones of the sun. It is approximately two-thirds of the distance to the center of the sun.

TAIL Characteristic of a comet that is visible when it is close enough to the sun, caused by evaporating material from the core of the comet.

TERMINATION SHOCK Point at which the speed of the solar wind slows abruptly due to interaction with the interstellar medium.

TIDAL FORCES Effects produced by the gravitational attraction of a body by another that can cause the deformation of the latter. An example is the moon's effect on Earth: the gravity of the moon "tugs" on the Earth's oceans, causing them to swell.

TIME-SPACE Mathematical model that combines the three dimensions of space with time, creating four dimensions.

TRANSIT Apparent movement of an object in front of another.

TRANSLATION Movement that changes the position of an object, different from rotation. It is commonly used to describe the change in position of a planet around a star, tracing an orbit around it.

TRIPLE-ALPHA PROCESS Series of reactions in which three helium nuclei are converted into one carbon nucleus. It only happens in very old stars.

TULLY–FISCHER RELATION Relation between the luminosity of a spiral galaxy and its angular speed. This allows for the calculation of the distances to spiral galaxies using their luminosity and apparent brightness. It is also one of the standard astronomical measurements.

U

ULTRAVIOLET One of the types of radiation on the electromagnetic spectrum. Its wavelength is shorter than visible light but longer than x-rays.

V

VELOCITY, ANGULAR Physical magnitude that is used to describe the movement of a body that rotates around an axis. The angular velocity depends on the angle that the body moves per unit of time.

VELOCITY, ESCAPE Minimum velocity needed to escape the gravitational influence of an object. The larger the object, the higher this velocity.

VELOCITY, RADIAL Physical magnitude used to describe when the distance between a body and a set point used as a reference changes per unit of time.

VORTEX Region in which the flow of a fluid rotates around an axis that can be straight or curved.

W

WAVELENGTH Distance between the repetitions of a wave or the separation between two identical regions of a wave, such as the separation between two crests.

WAVES, DENSITY Regions in the galactic disc in which the concentration of mass is higher than the rest. The spiral arms of a galaxy develop due to the presence of density waves.

WAVES, GRAVITATIONAL Deformation in the curvature of space-time caused by certain gravitational interactions. They propagate through space as waves at the speed of light.

WAVES, RADIO Type of electromagnetic radiation with a wavelength much larger than that of infrared. In astronomy, they are emitted by some of the most energetic objects in the cosmos.

INDEX

PHOTOGRAPHS AND ILLUSTRATIONS: 4–5: NASA/ESA/The Hubble Heritage Team (STScI/AURA); 6–7: SST/Institute for Solar Physics/Oddbjorn Engvold, Jun Elin Wiik, Luc Rouppe van der Voort; 8–9: NGP/Dana Berry; 10–11: Jordi Busqué; 14: NASA; 18–19: Mark A. Garlick; 20–21: Carles Javierre (Infographics); 22: (1) NASA/ESA/CXC/J. Strader (Michigan State University), (2) NASA/ESA/Hubble, (3) ESO, (4) ESA/Hubble & NASA; 23: NASA/ESA & The Hubble Heritage Team (STScI/AURA); 24–25 (UP): Juan Venegas; 24–25 (LO): Felipe García Mora; 26–27: ESO/S. Brunier; 26 (UP RT): Ken Crawford; 26 (LO CTR LE): Farmakopoulos Antonis; 26 (LO CTR RT): César Blanco González; 26 (LO RT): George Jacoby (NOAO) et al. & WIYN, AURA, NOAO, NSF; 27 (UP LE): NASA, ESA, N. Smith (Univ. California, Berkeley) et al. & The Hubble Heritage Team (STScI/AURA); 27 (UP RT): NASA/CXC/PSU/K. Getman, E. Feigelson, M. Kuhn & the MYStIX team & NASA/JPL-Caltech; 27 (LO): ESO/J. Emerson/VISTA & HLA, Hubble Heritage Team (STScI/AURA) & Robert Gendler; 28–29: Aleksandra Alekseeva/123rf; 30: ESA/NASA/JPL-Caltech;; 30–31: (1) Fermi & ROSAT, (2) IRAS/NASA & ESA Planck LFI & HFI Consortia, (3) ESA Planck LFI & HFI Consortia & Haslam et al., (4) WISE/NASA/JPL- Caltech/UCLA, (5) Haslam et al., (6) ROSAT & Nick Risinger; 32–33: Robert Gendler (robgendlerastropics.com); 34: NASA/JPL-Caltech/R. Hurt (SSC/Caltech);; 34–35: NASA/CXC/M. Weiss/Ohio State/A. Gupta et al.; 36–37: NASA/ESA/The Hubble SM4 ERO Team; 38 (UP): Mark A. Garlick; 38–39: Mark A. Garlick; 38 (LO): Leo Blitz/Carl Heiles/Evan Levine-UC Berkeley; 40–41: NASA; 41 (UP): ESA& The Hubble Heritage Team (STScI/AURA); 41 (LO): PdBI Arcsecond Whirlpool Survey; 42–43: Mark A. Garlick; 42 (UP): Felipe García Mora; 42 (LO): David Nidever (univ. Michigan & Virginia) & SDSS-III; 43 (LE): NASA/ESA& The Hubble Heritage Team STScI/AURA; 43 (CTR): Hubble Legacy Archive, NASA, ESA & Steve Cooper; 43 (RT): Adam Block/Mount Lemmon SkyCenter/University of Arizona; 44–45: David A. Aguilar (CfA); 48 (LE): D. D. Dixon (University of California, Riverside) & W. R. Purcell (Northwestern University);; 46–47: NASA/CXC/MIT/F. Baganoff, R. Shcherbakov et al.; 47: NASA/CXC/Stanford/I. Zhuravleva et al.; 48–49: Atlas Image [or Atlas Image mosaic] courtesy of MASS/Umass/IPAC-Caltech/NASA/NSF; 52–53: ESO/M. Kornmesser; 54: NASA/CXC/PSU/K. Getman et al. & NASA/JPL-Caltech/CfA/J. Wang et al.; 54–55 (UP): ESO/Digitized Sky Survey 2; 54–55 (LO): ESA/Herschel/PACS, SPIRE/Gould Belt survey Key Programme/Palmeirim et al.; 55 (UP): FORS Team/8.2-meter VLT Antu/ESO; 55 (LO): ESA/SPIRE/PACS/P. André (CEA Saclay); 56–57: NASA/JPL-Caltech/University of Arizona; 58–59: NASA; 58: G. Stinson (MPIA); 59 (1–6): NASA/ESA/C. Papovich (Texas A&M)/H. Ferguson (STScI)/S. Fabe; 60–63: Juan Venegas; 64: NASA/ESA/G. Bacon (STScI); 64–65: Felipe García Mora; 65: ESA/NASA; 68–69: Juan William Borrego Bustamante; 70–71: NASA/ESA/C.R. O'Dell (Vanderbilt University)/M. Meixner, P. McCullough, & G. Bacon (Space Telescope Science Institute); 72–73: Metagràfic; 74: Nick Risinger/Wikimedia Commons; 75: NASA/ESA/M. Robberto (Space Telescope Science Institute/ESA)/The Hubble Space Telescope; 76: (1) ESO/J. Emerson/VISTA, (2) NASA/ESA, (3) NASA/JPL-Caltech/L. Allen; 77: (4) NASA/ESA/M. Livio/The Hubble 20th Anniversary Team, (5) NASA/ESA/E. Sabbi (STScI), (6) ESO; 78–79: Juan William Borrego Bustamante; 80–81: Marc Reyes; 82–83: NASA/JPL-Caltech/Harvard- Smithsonian CfA; 84 (LO LE): Jean-Charles Cuillandre (CFHT)/Giovanni Anselmi (Coelum Astronomia)/Hawaiian Starlight; 84 (LO RT): NASA; 84–85: ESA/ESO; 85 (LO LE): ESA/Hubble/NASA; 85 (LO RT): ESO; 86–87: ALMA (ESO/NAOJ/NRAO)/M. Maercker et al.; 88–89: Juan Venegas; 90: NASA/ESA/The Hubble Heritage Team (STScI/AURA); 90–91 (UP): Carlos Milovic/Hubble Legacy Archive/NASA; 90–91 (LO): NASA/ESA/HEIC/The Hubble Heritage Team (STSCI/AURA); 91 (UP): Bill Snyder (BillSnyder Photography); 91 (LO): NASA; 92–93: NASA/JPL-Caltech; 94–95: NASA/ESA/J. Hester & A. Loll (Arizona State University); 96–97: Nathan Smith (University of California, Berkeley)/NASA; 97: ESA/NASA; 98: NASA/Don F. Figer (UCLA); 98 (LO LE): NASA/GSFC/Dana Berry; 99: ALMA (ESO/NAOJ/NRAO)/A. Angelich; 100: ALMA (ESO/NAOJ/NRAO)/A. Angelich; 100–101: NASA/CXC/MIT/L.Lopez et al.; 100–101: (infrared) Palomar, (radio) NSF/NRAO/VLA, (x-rays) NASA/CXC/University of Amsterdam/N.Rea et al., (optic) DSS; 101: (4) The Hubble Heritage Team (STScI/AURA)/Y. Chu (UIUC) et al./NASA, (5) NASA/CXC/SAO; 102: NASA/CXC/PSU/G.Pavlov et al.; 102–103: Juan Venegas; 106–107: Shutterstock/Anuchit kamsongmueang; 105: (x-rays) NASA/CXC/Caltech/P.Ogle et al (optic) NASA/STScI & R.Gendler (infrared) NASA/JPL-Caltech (radio) NSF/NRAO/VLA; 105 (INSET): Event Horizon Telescope Collaboration; 106–107: Mark A. Garlick; 107 (LO RT): ESO/L. Calçada; 108–109: Mark A. Garlick; 110: L. Chomiuk/B. Saxton/NRAO/AUI/NSF; 110–111: (x-rays) NASA/CXC/RIKEN/D.Takei et al. (optic) NASA/STScI (radio) NRAO/VLA; 111: NASA; 112: ESO; 112–113: NASA; 113 (UP): R. Hurt/Caltech-JPL; 113: (radio) NRAO/AUI/NSF, (infrared) JPL/Caltech, (visible) STScI, (ultraviolet) CXC, (x-rays) NASA/ESA; 116–117: Mark A. Garlick; 117 (UP): Felipe García Mora; 118–119: Mark A. Garlick; 119: Carles3Javierre (infographics); 120–121: Juan William Borrego Bustamante; 122–123: Felipe García Mora; 124–125: NASA; 126–127: NASA; 128 (UP): infographics; 128–129: SOHO, EIT Consortium, ESA, NASA; 130–131: TRAPPIST; 131 (UP LE): NASA's Scientific Visualization Studio/SDO Science Team/Virtual Solar Observatory; 131 (UP CTR): NASA/Howard Brown-Greaves; 131 (UP RT): Luc Viatour; 132–133: N. A. Sharp, NOAO/NSO/Kitt Peak FTS/AURA/NSF; 134–135 (UP): Felipe García Mora; 134–135 (LO): Juan William Borrego Bustamante; 135 (LO): Felipe García Mora; 136–137: ESA/Hubble; 138–139: ESO/L. Calçada, Nick Risinger; 139 (UP): NASA; 140: Carles Javierre (Infographics); 140–141: NASA/JPL-Caltech; 142–143: Mark A. Garlick; 144–145: NASA, ESA, Hubble SM4 ERO Team; 146–147: NASA's Goddard Space Flight Center; 148 (UP): Felipe García Mora; 148 (LO LE): NASA/SDO/AIA/LMSAL; 148 (LO RT): NASA's Goddard Space Flight Center/Duberstein; 148–149: NASA's Goddard Space Flight Center/SDO AIA Team; 150–151: NASA/IBEX/Adler Planetarium; 152 (LO): U.S. Air Force/Shawn Nickel; 152–153: NASA/ISS Expedition 23 crew; 153 (UP): NASA, ESA, Nichols (Univ. of Leicester); 153 (CTR): NASA/ESA/STScI/A. Schaller; 153 (LO): Carles Javierre (Infographics); 154 (UP): NASA/GSFC/SDO; 154 (LO): NASA/GSFC/SDO; 154–155: NASA's Goddard Space Flight Center/SDO/S. Wiessinger; 155: NASA/SDO; 156–157: NASA's Goddard Space Flight Center; 158–159: ESA; 159: NASA; 160–161: NASA; 161: Joan Pejoan; 162–163: WaterFrame/Alamy Stock Photo; 164–165: Carles Javierre (infographics); 168–169 (UP): Felipe García Mora; 168–169 (LO): Juan William Borrego Bustamante; 169 (LO): Felipe García Mora; 170–171: [Mercury] NASA/Johns Hopkins University Applied Physics Laboratory/Carnegie Institution of Washington; [Venus] NASA/JPL; [Earth] NASA; [Mars] NASA/JPL/Malin Space Science Systems; [Jupiter] Space Telescope Science Institute/NASA; [Saturn] NASA, ESA & Erich Karkoschka (Univ. of Arizona); [Uranus] NASA/JPL-Caltech; [Neptune] NASA/JPL; 172–173: Román García Mora; 174–175: Mark A. Garlick; 178–179: Eckhard Slawik/Science Photo Library; 182–183: NASA/ESA/Cassini Imaging Team; 182: NASA/JHUAPL/SwRI; 183: Felipe García Mora; 184: NASA; 185 (UP): NASA/JPL; 185 (LO): NASA/JPL; 186–187: NGP/Dana Berry; 188 (1) Archivo RBA, (2) Archivo RBA, (3) NASA, (4) ESA/MPAe, Lindau, (5) NASA/SDO, (6) NASA/JPL-Caltech/MSSS; 189 (1) Archivo RBA, (2) Getty Images, (3) NASA/JPL, (4) NASA/JPL/Univ. of Arizona, (5) NASA, ESA & J. Nichols (Univ. of Leicester), (6) NASA/JPL-Caltech/Space Science Institute, (7) ESA/Rosetta/NAVCAM, (8) NASA/Johns Hopkins University Applied Physics Laboratory/Southwest Research Institute, (9) NASA/JPL-Caltech/SwRI/MSSS/Gerald Eichstädt/Seán Doran; 190–191: Juan Venegas; 192–193: NASA/JPL-Caltech; 194–195: NASA/JPL-Caltech; 196–197: Mark Garlick; 196: Felipe García Mora; 198–199: NASA/JPL-Caltech/MSSS; 200–201: NASA/Johns Hopkins Applied Physics Laboratory/Arizona State University/Carnegie Science; 202 (UP): Jordi Busqué; 202 (LO): NASA/JPL; 203 (UP): NASA/JPL/Univ. of Arizona; 203 (LO): NASA/Johns Hopkins University Applied Physics Laboratory/Carnegie Institution of Washington; 204–205: Mark A. Garlick; 206–207: NASA/Johns Hopkins University Applied Physics Laboratory/Carnegie Institution of Washington; 207: NASA/JHUAP/Arizona State University; 208–209: NASA/JPL; 208: Mark A. Garlick; 210–211: NASA Earth Observatory/Robert Simmon image from Suomi NPP VIIRS/NOAA's Environmental Visualization Laboratory; 212–213: iStock/Helen Field; 214–215: NASA/JPL/USGS; 215: Trent Schindler/NASA; 216–217: NASA/JPL/Arizona State University, R. Luk; 216–217: José Saco (illustrations);; 218–219: MOLA Science Team/MSS/JPL/NASA; 220–221: NASA/JPL-Caltech/University of Arizona; 222–223: (Io) NASA/JPL/Univ. of Arizona, (Mars) NASA/Goddard Space Flight Center Scientific Visualization Studio & Virginia Butcher (SSAI); 222 (Idunn): NASA/JPL-Caltech/ESA; 223l: NASA; 223 (RT): NASA/JPL/Univ. of Arizona; 224–225: Mark A. Garlick;; 226–227: NASA/Johns Hopkins University Applied Physics Laboratory/Southwest Research Institute/Goddard Space Flight Center; 228: NASA/Goddard Space Flight Center; 229: (2a, 2b) NASA/JPL-Caltech/SSI, (3) NASA/JPL-Caltech, (4) NASA/JPL; 230–231: NASA/ESA; 230 (UP): NASA/JPL/TexasA&M/Cornell; 231: Felipe García Mora; 232–233: NASA/ESA/A. Simon (GSFC); 234–235: NASA/JPL-Caltech/SwRI/MSSS/Björn Jonsson; 235: ESO/Y. Beletsky; 236–237: NASA's Goddard Space Flight Center & Space Telescope Science Institute; 237bl: NASA/JPL/Space Science Institute; 237 (LO RT): NASA/JPL/Space Science Institute; 238–239: Mattias Malmer/Cassini Imaging Team (NASA); 240–241: NASA/JPL-Caltech/Space Science Institute; 242–243: NASA/JPL; 244–245: NASA/JPL-Caltech/SSI; 246–247: NASA/JPL-Caltech/SETI Institute; 246 (RT): Felipe García Mora; 247: (Enceladus) NASA/JPL/Space Science Institute, (Titan) NASA/JPL/Univ. of Arizona/Univ. of Idaho; 247: Felipe García Mora; 252–253: ESO/E. Slawik; 254–255: NASA/JPL; 254: NASA; 255 (LO): Román García Mora; 257 (LO): Román García Mora; 258: NASA/JHUAPL/Swri; 258–259 (2): NASA/JPL-Caltech/UCLA/MPS/DLR/IDA; 259 (3, 4): NASA/JPL-Caltech/R. Hurt (SSC-Caltech); 260–261: Felipe García Mora; 260 (LO): ESA/Rosetta/NAVCAM; 262–263: Juan Venegas; 266–267: NASA/FUSE/Lynette Cook; 267: NASA/R. Hurt/T. Pyle; 268–269: IAU/L. Calçada; 270–271: Felipe García Mora; 273: NASA/JPL-Caltech/R. Hurt (SSC/Caltech); 274–275 (UP): NASA/JPL-Caltech; 274–275 (LO): Carles Javierre (infographics); 276–277: Carles Javierre (infographics); 278–279: Danielle Futselaar & Franck Marchis/SETI Institute; 279 (UP): NASA/Goddard/Francis Reddy; 279 (CTR): NASA/JPL-Caltech; 279 (LO LE): The Mars Underground; 279 (LO RT): The Mars Underground; 280–281: NASA/JPL-Caltech/R. Hurt (IPAC); 281 (LO): PHL (UPR Arecibo); 282–283: ESO/M. Kornmesser; 283 (LO): Archivo RBA; 284: (1) PHL/UPR Arecibo/NASA Epic Team, (2) PHL/UPR Arecibo/ESO/S. Brunier, (3) NASA/JPL-Caltech; 285: (1) NASA Ames/JPL-Caltech/T. Pyle, (2) NASA Ames/SETI Institute/JPL-Caltech, (3) ESO/M. Kornmesser; 286–287: ESO/M. Kornmesser/Nick Risinger; 287: Royal Observatory Edinburgh/Anglo-Australian Observatory/AURA; 288–289: ESA; 289: NASA/JPL-Caltech/H. Knutson (Harvard-Smithsonian CfA); 292–293: Felipe García Mora; 293 (UP): NASA/ESA/The Hubble Heritage Team (STScI/AURA)-Hubble/Europe Collaboration; 293 (UP RT): ESA/ATG medialab/C. Carreau; 293 (LO): NASA/ESA/The Hubble Heritage Team (STScI/AURA); 294–295 (UP): Sloan Digital Sky Survey (SDSS); 294–295 (UP): Felipe García Mora; 296–297: Felipe García Mora; 298–299: Felipe García Mora; 298 (LO): R. Brent Tully et al., Nature Publishing Group; 300 (LO LE): NASA/Swift & Bill Schoening, Vanessa Harvey/REU program/NOAO/AURA/NSF; 300 (LO RT): NASA/JPL-Caltech/K. Gordon (Univ. Arizona); 300 (LO LE): ESA/Herschel/PACS/SPIRE/J.Fritz, U. Gent/XMM-Newton/EPIC/W. Pietsch, MPE; 301 (LO RT): UV-NASA/Swift/Stefan Immler (GSFC) & Erin Grand (UMCP), Bill Schoening, Vanessa Harvey/REU program/NOAO/AURA/NSF; 301 (UP LE): NASA, ESA & the Hubble Heritage Team (STScI/AURA)-ESA/Hubble Collaboration; 301 (UP RT): Fernando de Gorocica; 302 (background): ESO/R. Gendler; 302 (UP LE): ESA/Hubble & NASA; 302 (UP RT): ESO/L. Calçada; 302 (LO LE): ESO; 302 (LO RT): NASA, ESA, F. Paresce (INAF-IASF, Bolonia), R. O'Connell (Univ. of Virginia, Charlottesville) & the Wide Field Camera 3 Science Oversight Committee; 303 (background): ESA/Hubble & Digitized Sky Survey 2; 303 (UP): David L. Nidever, et al., NRAO/AUI/NSF & Mellinger, Leiden/Argentine/Bonn Survey, Parkes Observatory, Westerbork Observatory, Arecibo Observatory; 303 (CTR RT): ESA/NASA; 303 (CTR LE): A. Nota (ESA/STScI) et al., ESA, NASA; 303 (LO): NASA, ESA & the Hubble Heritage Team (STScI/AURA)-ESA/Hubble Collaboration; 304–305: (1) NASA; (2) Astrodon; (3) P. Massey/Lowell Observatory & K. Olsen/NOAO/AURA/NSF; (4) Local Group Galaxies Survey Team/NOAO/AURA/NSF; (5) Fabrizio Francione; (6) ESO; (7) Robert Gendler, Subaru Telescope, National Astronomical Observatory of Japan (NAOJ); 306: Felipe García Mora; 306–307: NASA; 306–307: (1) NASA, (2) ESA, (3, 4) The Hubble Heritage Team (STScI/AURA)-ESA/Hubble Collaboration & A. Evans (Univ. Virginia, Charlottesville/NRAO/Stony Brook University), (5) K. Noll (STScI), (6) J. Westphal (Caltech); 308–309: NASA/STScI; 323: NASA/CXC/JPL-Caltech/STScI.

VISUAL GALAXY

Since 1888, the National Geographic Society has funded more than 13,000 research, exploration, and preservation projects around the world. National Geographic Partners distributes a portion of the funds it receives from your purchase to National Geographic Society to support programs including the conservation of animals and their habitats.

National Geographic Partners
1145 17th Street NW
Washington, DC 20036-4688 USA

Get closer to National Geographic explorers and photographers, and connect with our global community. Join us today at nationalgeographic.com/join

For information about special discounts for bulk purchases, please contact National Geographic Books Special Sales: specialsales@natgeo.com

For rights or permissions inquiries, please contact National Geographic Books Subsidiary Rights: bookrights@natgeo.com

ISBN: 978-1-4262-2060-9

Text: Joan A. Catalá, Joel Gabàs, Jordi L. Gutiérrez, Alba Llabrés, Jordi Pereyra, Alejandro Riveiro de la Peña, Rosa Rodríguez Gasén
Complementary texts and appendices: Luz María Bazaldúa, Gonzalo del Castillo, Alejandro Riveiro de la Peña, Juan Romero, José Saco

Special thanks to: Bridget Hamilton, Moriah Petty, Richard Rothschild, Laura Daly, Kate Armstrong, Diane Ersepke, and Andrew Fazekas

Printed in China

19/PPS/1

The Small Magellanic Cloud (SMC) galaxy, cousin to the Large Magellanic Cloud, is one of the Milky Way's closest neighbors. Although it is a small galaxy and 200,000 light-years away, the SMC is so bright that it is visible to the naked eye from the Southern Hemisphere and near the equator. Many navigators, including Ferdinand Magellan who lends his name to the SMC, used it to help find their way across the oceans.

NAVIGATE THE UNIVERSE

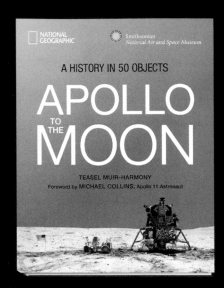

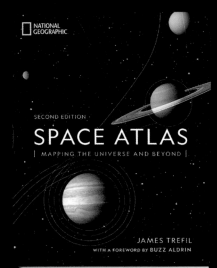

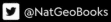